科技翻译教程

主　编◎王兴龙　曹顺发
副主编◎余　胜　张钺奇　屈宇昕

人民交通出版社股份有限公司
北　京

内 容 提 要

本教材以科技文体的特征、科技翻译的原则与标准、译者素养、网络查询工具与方法等从事科技翻译工作所需的基本知识与技能为切入点，通过英汉两种语言中科技文体的特点对比，详细介绍了科技翻译中常见的翻译方法和技巧。本教材注重翻译理论与实践的有机融合，选取具有针对性和实用性的语料，通过机电工程、交通运输工程、土木工程和材料工程等领域的译例分析，详细介绍了科技翻译实践中的常见问题和解决方案。

本教材实用性较强，推荐作为高等院校英语（翻译）专业高年级本科生及翻译硕士科技翻译课程教材使用，也可用作非英语专业本科生及研究生科技翻译类选修课程教材。

图书在版编目(CIP)数据

科技翻译教程/王兴龙，曹顺发主编. —北京：人民交通出版社股份有限公司，2023.8

ISBN 978-7-114-18862-6

Ⅰ.①科… Ⅱ.①王… ②曹… Ⅲ.①科学技术—英语—翻译—教材 Ⅳ.①G301

中国国家版本馆 CIP 数据核字(2023)第 116229 号

Keji Fanyi Jiaocheng

书　　名：科技翻译教程

著 作 者：王兴龙　曹顺发

责任编辑：齐黄柏盈

责任校对：赵媛媛　龙　雪

责任印制：张　凯

出版发行：人民交通出版社股份有限公司

地　　址：(100011)北京市朝阳区安定门外外馆斜街 3 号

网　　址：http://www.ccpcl.com.cn

销售电话：(010)59757973

总 经 销：人民交通出版社股份有限公司发行部

经　　销：各地新华书店

印　　刷：北京虎彩文化传播有限公司

开　　本：787×1092　1/16

印　　张：14.75

字　　数：306 千

版　　次：2023 年 8 月　第 1 版

印　　次：2023 年 8 月　第 1 次印刷

书　　号：ISBN 978-7-114-18862-6

定　　价：48.00 元

FOREWORD | 前言

在长期的科技翻译教学过程中,我们发现尽管科技翻译教材种类繁多,但多偏重于一般英汉或汉英翻译理论及方法的介绍,与真正的“科技”文体联系不紧密。同时,现有教材对学生翻译实践能力的培养不够重视,忽视学生未来工作中职业素养的养成问题。为此,我们组织了几位从事科技翻译教学的一线教师及翻译研究的专家和学者,编写了这本《科技翻译教程》。

本教材的编写紧跟翻译学科及科技发展前沿,注重翻译理论与翻译实践的有机融合,翻译语料具有针对性和实用性,注重译例分析,每一章后都配有针对性练习,并附有参考答案。本教材实用性较强,针对高等院校英语(翻译)专业高年级本科生及翻译硕士科技翻译课程编写,也可用作非英语专业本科生及研究生科技翻译类选修课程教材。

全书共四个部分,分为16章。

第一部分为科技翻译入门,涉及科技文体的特征、科技翻译的原则与标准、译者素养、网络查询工具与方法等从事科技翻译工作所需的基本知识与技能。

第二部分为科技文体的解构与重构,对比英汉两种语言中科技文体的特点,着重讲解增译与省译、拆译与合译、词性转换、反译等科技文体中的翻译方法。

第三部分为科技翻译中的常见表达,涉及数字的翻译、主动式与被动式的翻译、从句的翻译、特殊结构的翻译等科技翻译中尤其需要注意的问题。

第四部分为科技语篇译例分析,通过机电工程、交通运输工程、土木工程和材料工程等领域的语篇译例分析,详细介绍了科技翻译实践中常见的问题及解决方案。

西南大学外国语学院董子畅、李鑫、王婷作为编者参与了本教材的编写。同时,因本教材涉及学科知识较广,在编写过程中,张青松博士、余强工程师对土木工程、材料工程、交通运输工程等领域专业知识进行了解答。熊伟教授、林隆棋副教授、Steven Gauntt 博士及 Daniel Barkley 博士为本教材的译文审校提供了语言支持,谨向以上人员表达诚挚的谢意。本教材是重庆交通大学规划教材建设项目的成果,在编写过程中得到了重庆交通大学各部门的大力支持和帮助,在此表示衷心感谢。

本教材从内容到形式都有不少新的尝试与探索,但由于编者水平有限,不足之处在所难免,敬请各位专家、老师和同学不吝指正,以使教材日臻完善。

编　者

2023年5月

CONTENTS | 目录

第一部分

科技翻译入门

第二部分

科技文体的解构与重构

第三部分

科技翻译中的常见表达

第四部分

科技语篇译例分析

第一部分

科技翻译入门

第一章 科技文体的特征

文体,指独立成篇的文本体裁。根据体裁特征,语篇可以分为文学文体、政治文体、法律文体、新闻文体、科技文体等。不同的文体各有其特点,如文学文体以艺术性、感染性和美学性为特点,政治文体以宣传性、动员性和鼓舞性为特点,法律文体以法律性、权威性和严谨性为特点,新闻文体以新颖性、时效性和简洁性为特点……因此,在翻译时不但需要考虑语言转换、信息转换、文化转换,还需要考虑原文的文体特征,做好文体转换。

科技翻译在遵循翻译的基本要求的基础上,还需要充分了解科技文体的特征,力求翻译得“原汁原味”。在分析科技文体的特征之前需要明确什么是科技文体。不同于其他文体,科技文体主要指谈及科学和技术的书面语和口语,其内容涉及科学、技术、工程等专业领域,具体包括:①科技著作、科技论文和报告、实验报告和方案;②各类科技情报和文字资料;③科技使用手册,包括仪器、仪表、机械、工具等的结构描述和操作规程;④有关科技问题的会谈、会议、交谈的用语;⑤有关科技的影片、录像等有声资料的解说词,等等。以科学、技术、工程等专业知识为基础的科技文体,总的来说主要有以下五大特征:

(1)专业性:相比于其他文体,科技文体表达的内容更加专业,较多使用专业术语、符号、公式、图表等。

(2)客观性:科技文体要求客观地描述相应事实,不能在论述中带有作者的主观偏见。

(3)精确性:科技文体要求精确地描述相应事实,不能有模糊不清、似是而非的表述。

(4)细节性:科技文体要求提供详细的信息,以便读者能更好地理解和验证文本的内容。

(5)逻辑性:科技文体要求文章的内容条分缕析,逻辑严密,层次分明,以便读者能够准确无误地理解文章内容。

了解科技文体的特征,有利于提高文本翻译的质量,一方面可以更加专业、准确地引进国外先进的科学技术知识,另一方面也能够更好地传播我国优秀科研成果,提升我国的科技影响力,推动科技发展进步。

除专业性、客观性、精确性、细节性和逻辑性五大总体特征之外,科技文体在词汇和句法层面也有相应的特征,下面将专门对其展开论述。

第一节 词汇特征

一 专业性

与科技文体特征一脉相承,科技文体中的词具有专业性特征。科技文体中的词汇根据专业程度可分为专业词汇(technical word)、半专业词汇(sub-technical word)和普通词汇三类。专业词汇指专门用于科技文体中的词语,多是源于科学技术发展带来的新兴事物和发现,如“ultrasonic(超声波)、entropy(熵)、bandwidth(频带宽度)、anode(阳极)、hydroxide(氢氧化物)”等。半专业词汇指由普通词经意义扩展形成的具有专业意义的词语,如“script”一词源于拉丁词“scriptum”,原义为“手写的文件”,后应用于文学领域指“剧本、讲稿”,应用于语言学领域指“(一种语言的)字母系统、字母表”,应用于计算机领域指“脚本”,应用于医学领域指“处方”等。与之类似的还有“cache”,原义为“隐藏物(如武器)”,应用于计算机领域指“高速缓冲存储器”;“function”,原义为“功能”,应用于生物学领域指“机能”,应用于数学领域指“函数”;“nucleus”,原义为“核心、中心”,应用于生物学领域指“细胞核”,应用于物理学、化学领域指“原子核”。普通词汇即指英语中的常用词语,也是科技英语中使用频率最高的词。科技文体中专业词汇和半专业词汇的使用使得文章更具专业性,要求讲述者及作者具备该领域的专业知识素养,同时也要求听众和读者具备相应的专业词汇储备。

值得注意的是,语言是一个不断变化、发展的系统,语言系统内的词同样处于不断变化、发展之中。随着社会、科技、文化等的不断发展,新的词语逐渐形成。但由于语言具有经济性,词的数量不会无限制地增长,于是许多普通词产生了新的意义,从而变成所谓的半专业词。与普通词可以逐渐演变成半专业词相似,一些专业词语在深入大众生活之后专业性也会逐渐减弱,直至融入日常生活之中。如“electricity”起初也是物理学领域内的专业词语,但随着电的广泛应用,该词之意可谓无人不知。

此外,科技文体中倾向于使用较为正式的动词,从而彰显其专业性,比如,比起“take in”更倾向于使用“absorb”,比起“put together”更倾向于使用“aggregate”,比起“come out”更倾向于使用“emerge”,比起“make sure”更倾向于使用“ensure”,比起“give”更倾向于使用“supply”,比起“enough”更倾向于使用“sufficient”,等等。

二 单义性

科技文体中词汇的另一特征是其单义性。由于科技文体对于精确性的要求,为避免歧义,词汇总体呈单义性。这与上文提到的半专业词汇是否相矛盾?答案自然是否定的。半专业词汇,如“script,cache,function,nucleus”等虽有多种意义,但在每个学科领域内都保持各自的单义性特点。以“power”一词为例,其原义为“能力”,如例(1)所示。但在不同学科、不同领域中,“power”又有着其他意义,如在电力学中指“电力”,见例(2);在物理学中指“功率”,见例(3);在数学中指“乘方,幂”,见例(4),等等。尽管该词存在多种含义,但在不同的领域内各有其对应的意义。

(1) The real beauty of the Java language, however, is its power to save users' money, because it vastly simplifies creating and deploying applications and because it lets them keep their existing “legacy” computers and software.

译文 然而,Java 语言的真正优点是具有让用户省钱的能力,因为它大大简化了应用程序的开发与部署使用,而且让它们保留已有的“传统”计算机和软件。

(2) Redundant power supplies, hot-swappable modules and Spanning Tree Protocol (STP) support are all mandatory elements of a backbone switch, as is support for emerging switch technologies, including virtual LANs.

译文 具有冗余电力的电源、可热插拔的模块和对生成树协议(STP)的支持都是基干交换机必备的成分,实际上它支持新出现的交换技术,如虚拟局域网。

(3) This powerful electric juicer automatically separates the pulp from the juice.

译文 这种大功率的电动榨汁机可自动分离果肉和果汁。

(4) A logarithm tells you what power a base number must be raised to in order to produce a given value.

译文 对数说明一个底数的几次幂等于一个给定的值。

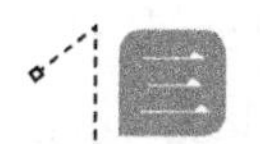

三 简略性

科技文体客观、准确、简明,有逻辑性和层次性。科技文体中词汇的特征之一即是简略性,化繁为简,简明清晰。这一点主要体现在科技文体中广泛使用的截短词(clipping)和缩略词(abbreviation)。截短词,即裁剪较长的单词,选取其中某部分以代替该词,如“flu”是由“influenza”裁剪而来,“modem”由“modulator-demodulator”裁剪而来,“fax”由“facsimile”裁剪而来,“copter”由“helicopter”裁剪而来,等等。将较长词组中每个单词首字母提取出来,可形成首字母缩略词。根据发音方式,可以分为按词发音的首字母缩略词(acronym)和按字母发音的首字母缩略词(initialism)两类。按词发音的首字母缩略词即指提取出来的缩略词可以像单词一样单独发音,如 NASA (National Aeronautics and Space Administration)、AIDS (Acquired Immune Deficiency Syndrome)、maser(microwave amplification by stimulated emission of radiation)等。按字母发音的首字母缩略词则是指提取出来的缩略词按词中的字母逐一发音,如CPU(Central Processing Unit)、HDD(Hard Disk Drive)、USB(Universal Serial Bus)、HTML(Hypertext Markup Language)、CSS(Cascading Style Sheets)、SQL(Structured Query Language)、FTP(File Transfer Protocol),等等。如示例所示,科技文体中时常使用这些具有简略性特征的截短词和缩略词,从而使语言更加简洁精练。

A million miles from the earth, NASA's Advanced Composition Explorer (ACE) spacecraft suddenly found itself under assault.

译文 在离地球大约100万英里(约1609344千米)的地方,美国国家航空航天局(NASA)的航天器——高级太空成分探索者(ACE)突然遭受攻击。

值得注意的是,同一缩略词在不同学科领域内通常表示不同的意思,如"GPA"既可以指"通用放大器"(General Purpose Amplifier),也可以指"客运总经纪人"(General Procurement Agent),还可以指"平均分数"(Grade-Point Average)。因此,结合文本信息与相关专业知识,正确把握每个缩略词所代表的意义颇为重要。

四 生成性

科技文体中的专业词汇及半专业词汇并不是凭空产生的,大多是由已有词、词根和词缀通过一定的构词法形成的,因此具有生成性。以下主要介绍科技文体中通过相关构词法形成的三类词,分别为复合词、拼缀词与派生词。

(一)复合词

复合词指通过复合法(compounding)形成的词,即由两个或两个以上自由词组成的词。科技文体中的许多词都是通过复合法形成的复合词,主要包括以下三种形式:①由连字符连接的复合词,如user-friendly(user + friendly)、real-time(real + time)、cloud-based(cloud + based)、data-driven(data + driven)等;②无连字符但有空格的复合词,如cloud computing(cloud + computing)、artificial intelligence(artificial + intelligence)、machine learning(machine + learning)、big data(big + data)等;③无连字符且无空格的复合词,如software(soft + ware)、hardware(hard + ware)、firmware(firm + ware)、cybersecurity(cyber + security)、blockchain(block + chain)等。

(二)拼缀词

拼缀词指通过拼缀法(blending)形成的词。与复合词不同的是,拼缀词由两个或以上词的部分拼凑组成,而非两个或以上词的直接拼凑。拼缀词可以是第一个词的整体与第二个词的某部分拼凑,如"webinar"是由"web"与"seminar"中的"inar"拼凑而成的;也可以是由第一个词的部分与第二个词的某部分拼凑的,如"technophile"是由"technology"中的"techno"与"philosophy"中"phy"的变体拼凑而成的,"infobahn"是由"information"中的"info"和

"autobahn"中的"bahn"拼凑而成的;还可以由第一个词的某部分与第二个词的整体拼凑,如"technobabble"是由"technology"中的"techno"和"babble"拼凑而成的。

(三)派生词

派生词指通过派生法(derivation)形成的词,即在词根前面加前缀或在词根后面加后缀构成的新的词语。比如在词根前加前缀"multi-"(多)形成的 multi-platform(多平台的)、multi-core(多核的、多芯的)、multi-media(多媒体)、multi-user(多用户的、多人联机的)、multi-tasking(多任务的)等,以及在词根后面加后缀"-ify"构成动词(使……化),比如 purify(净化)、quantify(量化)、simplify(简化)等。

第二节 句法特征

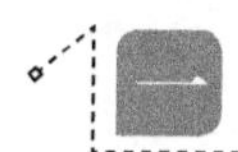

多使用名词化结构

名词化是指将句子中的动词、形容词等通过一定方式(如加缀、转化等)转换为名词的语法过程。英语科技文体中较多使用名词化结构,使得语言更加精练、客观和正式,并能更好地概括所描述事件的全貌。需要注意的是,由于汉语是一种偏动态的语言,即惯于使用动词,因此,在科技文体中使用名词化结构的情况明显少于英语。

(1) Twenty years earlier, television and satellite technology helped play a role in the fall of the Wall, by connecting people and empowering them with information.

译文一 二十年前,电视和卫星技术用所提供的信息将人们联系在了一起,为推倒柏林墙贡献了力量。

译文二 二十年前,电视和卫星技术用所提供的信息将人们联系在了一起,为柏林墙的倒塌贡献了力量。

分析 译文一选取了动宾结构翻译"the fall of the Wall",而译文二选取了名词化结构。两个译文看似意义相近,但在识解同一事件的方式上有所差别,前者采取的是渐次扫描(sequential scanning)的心理扫描方式,而后者则采取了总括扫描(summary scanning)的心理扫描方式。相较而言,后者比前者更忠实于原文,更能还原该事件的全貌。

(2) High blood pressure is a contraindication for this drug.

译文一 高血压是这种药物的禁忌症。

译文二 高血压者忌服此药。

分析 译文一是原文的对应翻译,将"a contraindication for this drug"直接译为"这种药物的禁忌症”,读者一读便可感受到该译文有悖于汉语的表达习惯。译文二将这一名词化结构转换为汉语常用的动宾结构“忌服此药”,明显更符合汉语思维和表达习惯。

二 多使用非谓语动词

科技文体中常用非谓语动词来替代复杂的从句,从而简化句子结构,使得语言更加简明清晰、结构更加紧凑、逻辑更加清楚。科技文体中的非谓语动词主要包括不定式、动名词以及(现在和过去)分词。

(1) Tornadoes, the deadliest weather disaster to hit the country this year, present a particularly thorny case. (不定式)

译文 今年袭击该国最致命的气候现象飓风,就是一个尤其棘手的例证。

(2) Determining your foot type is key to making sure you get the right running shoes. (动名词)

译文 确定脚型才是购买合适跑鞋的关键。

(3) By extension, aloha connects humans with the natural world, putting individual humans into context with animals, plants, and the Earth. (现在分词)

译文 推而广之,“阿罗哈”使人类与自然世界联系起来,让每个人与动物、植物和地球之间产生共鸣。

(4) Informally known as Mount Sharp, in honor of Robert Sharp, a pioneering planetary scientist at Caltech, it is taller than any mountain in the continental United States. (过去分词)

译文 这座山高于美国本土的任何一座,民间称之为夏普山,以纪念世界太空探险先驱,即加州理工学院的行星科学家罗伯特·夏普。

三 多使用后置定语

在阐明科学技术时,科技文体需对其进行明确的限定,以增强文本的精确性,因此常采用后置定语,一方面可以使文本之间的逻辑关系显得更加清楚、层次分明,另一方面则有助于加强内容之间的连贯性。

(1) ChatGPT is a new artificial intelligence (AI) technology that enables natural language conversations between humans and machines.

译文 ChatGPT 是一项新的人工智能(AI)技术,让人机自然语言对话成为现实。

(2) Some of ATM's benefits include its cell-based switching scheme that protects traffic from

the latency problems it faces over shared media.

译文 异步传输模式(ATM)的优势包括基于信元的交换方案保护信息流,从而避免共享媒体中的延迟问题。

四 多使用长句

科技文体涉及科学与技术相关内容,因此需要做到客观、精确、严谨、逻辑清晰、层次分明。为使所描述的现象或概念清晰准确,科技文体中多采用长句,一层套一层,将所述内容阐述清楚,避免歧义。

(1) What we do know now is that although hydrophobic down does manages a certain degree of moisture better than standard down, it is not an equal substitute for synthetic insulations when soaking wet — which is something critically important to think about if you're engaging in activities in which you might get wet.

译文 就人们目前所知而言,拒水羽绒虽然比普通羽绒更能控制潮气,但当完全湿透后,却不能替代人造棉(合成纤维类人造保暖材料)。使用者若要从事会沾水的活动,考虑到这一点则至关重要。

(2) The pose of the moving platform relative to the base is thus defined by a position vector p in addition to a rotation matrix $\boldsymbol{R}$, in which p denotes the position vector of the origin of $\{B\}$ with respect to frame $\{A\}$, and furthermore, the orientation of $\{B\}$ with respect to $\{A\}$ is represented by a 3×3 rotation matrix $\boldsymbol{R}$.

译文 运动平台相对于基座的姿态是这样定义的,通过一个位置向量 p 和一个旋转矩阵 $\boldsymbol{R}$,p 表示 $\{B\}$ 的原点相对于坐标系 $\{A\}$ 的位置向量。另外,$\{B\}$ 相对于 $\{A\}$ 的方位通过一个 3×3 的旋转矩阵 $\boldsymbol{R}$ 来表示。

五 多使用现在时态

科技文体中所使用的时态主要为一般现在时、现在完成时、一般过去时和一般将来时四种,其中一般现在时(用以描述客观真理、自然现象、常规事件等)最为常见。现在完成时、一般过去时主要是对过去研究进行描述时所使用的时态,一般将来时则主要是对未来研究进行展望时所使用的时态。

(1) The field of artificial intelligence, or AI, goes further still: it attempts not just to understand but also to build intelligent entities.

译文 人工智能(AI)领域则更进一步:它不仅试图理解,而且还试图建造智能实体。

(2) ATM represents a radical departure from the traditional LAN format.

译文 采用异步传输模式(ATM)意味着从传统局域网格式的彻底脱离。

练习题

一、英译汉

1. Network administrators also gain power and control through the ability to directly access and manipulate configuration database files.

2. Because a USB port supplies power to peripherals, it will weed out the tangle of power cords and space-hogging power transformers.

3. In addition, DTR concentrators can be linked to each other over a LAN or WAN via data transfer services such as Asynchronous Transfer Mode.

4. The hills are remnants of an earlier geological era, scraped bare of most soil and exposed to the elements.

5. NASS acts as a specialized "data mover", pushing and pulling files such as text, images and video clips over a network.

6. After annealing, samples were cooled down within approximately thirty seconds to room temperature.

7. An ELF wave can transmit only a few bits of information per second.

8. The existing system could not ensure automatic control of increased acid concentration.

9. Cast iron should not be thought of as a metal containing a single element, but rather, as one having in its composition at least six elements.

10. It is one of the stars visible to the naked eyes.

二、汉译英

1. 3 的 3 次方是 27。

2. 每个虚拟线路上的通信可以用其特定的密钥进行加密。

3. 关键是每家供应商都有单一的、得到很多改进的且定价更具有吸引力的产品集。

4. 远程访问一个公司的内部网可通过多种方式实现。

5. 钻石是碳元素的晶体结构。

6. 几乎所有的金属都是良导体,银更是其中最好的。

7. 热胀冷缩是一切物质的共性。

8. 玻璃不导电,空气也不导电。

9. 鲸鱼和鲨鱼的不同之处在于,鲸鱼是哺乳类动物,而鲨鱼是鱼类。

10. 电流变化时,磁场也会发生变化。

第二章
科技翻译的原则与标准

在讨论科技翻译的原则与标准之前，需首先了解翻译的原则与标准，即在翻译中所应当遵循的相关准则。这些准则不仅有助于指导翻译实践，还可以用于评估译文质量。

关于翻译原则与标准的研究早已有之。我国古代译入佛经时主张直译，如释道安提出的"五失本"和"三不易"原则，认为译者应按照原文不加改动地直接翻译。而鸠摩罗什则主张意译，注重翻译的可读性。至唐朝时，玄奘提出"新译"，认为"既须求真，又须喻俗"，即在真实反映原文的基础上使译文更加通俗易懂。到了近代，严复提出的"信、达、雅"可谓是最广为人知的翻译原则与标准。其中，"信"指忠实于原文，"达"指语言表达的流畅，"雅"指译文的尔雅。刘重德在严复"信、达、雅"翻译标准的基础上，提出了"信、达、切"的翻译标准。他认为"雅"这一标准并不合适，并非所有文体都具有"雅"这一特质，从而主张将"雅"改为"切"，要求译文切合原文文体风格。

新中国成立后，关于翻译原则与标准的观点更是如雨后春笋般出现，如傅雷的"神似观"、钱钟书的"化境说"、许渊冲的"三美论"、辜正坤的"翻译标准多元互补论"等。傅雷在《高老头》重译本序中明确指出："以效果而论，翻译应当像临画一样，所求的不在形似而在神似……各种文学各有特色，各有无可模仿的优点，各有无法补救的缺陷，同时又各有不能侵犯的戒律。像英、法，英、德那样接近的语言，尚且有许多难以互译的地方；中西文字的扞格远过于此，要求传神达意，铢两悉称，自非死抓字典，按照原文句法拼凑堆砌所能济事。"自此，"重神似而不重形似"的翻译观点便广泛流传开来，产生了深远的影响。钱钟书在《林纾的翻译》中指出："文学翻译的最高理想可以说是'化'。把作品从一国文字转变成另一国文字，既能不因语文习惯的差异而露出生硬牵强的痕迹，又能完全保存原作的风味，那就算得入于'化境'……译本对原作应该忠实得以至于读起来不像译本，因为作品在原文里决不会读起来像翻译出的东西。"钱钟书的"化境说"与傅雷的"神似观"有一定的相通之处，都强调还原原文的"魂"。许渊冲将鲁迅谈论文章写作时提出的"意美以感心，一也；音美以感耳，二也；形美以感目，三也"应用于翻译中，认为翻译应满足"意美、音美、形美"的要求。三者之中，"意美"为主，"音美"其次，"形美"再次。因此，当三者不可兼得时，以"意美"为首选。辜正坤提出了"翻译标准多元互补论"，认为翻译的标准不止一个，具有多元性。翻译标准的多元化并不是翻译标准的全元化，也不是翻译标准的虚无化，而是追求无限中的有限性。他认为多元化翻译标准是一个由若干标准组成的相辅相成的标准系统，其中包括绝对标准、最高标准和具体标准。翻译的绝对标准指原作本身,翻译的最高标准是最佳近似度,而具体标准不止一个。其中,绝对标准可以看作是最高标准的标准,最高标准可以看作是具体标准的标准。各个具体标准又有主次之分,但由于时间、空间以及认识主体(人)的种种因素,主次标准可能发生变化,主标准可能会降为次标准,次标准也可能会升为主标准,也可能会有新的标准产生。同时,各标准之间具有互补性,相辅相成。

国外也有许多关于翻译原则与标准的探讨。Tytler 在《论翻译的原则》(*Essay on the Principles of Translation*)一书中表明,好的翻译是将原文的价值完全注入译文,使译文读者能够充分理解原文,产生与原文读者读原文相同的感受。他提出翻译的三原则,一是翻译应完全复写出原文思想,二是译文的风格和笔调应与原文的性质相同,三是译文应和原文一样流畅。Jakobson 在《论翻译的语言问题》(*On Linguistics Aspects of Translation*)一书中提出"翻译对等"的概念,认为译文与原文应实现信息对等,即内容对等。翻译对等这一原则与标准对西方翻译界影响深远,翻译对等究竟指的是什么对等,不同的学者有不同的看法。Catford 在《翻译的语言学理论》(*A Linguistic Theory of Translation*)一书中提出了"形式对等"的翻译原则与标准,即译文中的范畴成分在目的语中和原文中的范畴成分在源语中相同,并认为必要时可进行范畴转换。Nida 早期也聚焦信息本身,指出翻译应寻求形式对等;后期他又提出了功能对等理论,即动态对等,认为翻译对等最重要的是读者之间的反应对等而非形式对等,即译文读者理解和接受译文的程度和原文读者理解和接受原文程度的对等。Newmark 认为翻译实质上是一种意义的转移,并在他的著作《翻译问题探讨》(*Approaches to Translation*)中提出"语义翻译"与"交际翻译"两个概念,前者更忠实于原文,强调译文与原文的意义对等,而后者更注重读者反应,力求使译文读者获得与原文读者相同的阅读体验。Koller 指出译文与原文的对等应考虑多种因素:一是文本传达的言外内容,即外延对等;二是措辞方式传达的内涵意义,即内涵对等;三是特定文本类型的文本和语言规范,即文本规范对等;四是翻译指向接受者,即语用对等;五是源语文本的形式与美学特征,即形式对等。

由是观之,国内外对于翻译原则与标准的研究成果丰硕,虽然目前学界对此尚未形成统一的观点,但其中有一些共同之处,如大多学者都强调翻译的忠实和流畅。翻译的原则与标准都需遵守,但不同文体由于其特性不同,翻译的侧重点亦有所不同。比如,文学文体翻译时侧重艺术性、感染性和美学性,政治文体翻译时侧重宣传性、动员性和鼓舞性;法律文体翻译时侧重法律性、权威性和严谨性;新闻文体翻译时侧重新颖性、时效性和简洁性。

第一章中我们讨论了科技文体的五大总体特征,即专业性、客观性、精确性、细节性和逻辑性。鉴于科技文体的特殊性,在遵循翻译的原则与标准的基础上,又有其特殊要求。了解与掌握科技翻译的原则与标准,对于科技翻译实践以及提高翻译质量有着极为重要的作用。针对科技文体中的上述特征,本书将科技翻译的原则与标准总结为:忠实、通顺、准确、简洁。

第一节 忠 实

科技翻译首先需要做到的无疑是对原文的忠实,这也是翻译的基本原则与标准。忠实于原文需要译者具备专业性与客观性。专业性要求译者具备专业知识,在翻译科技文体时做到不误译;客观性要求译者具备专业能力,在翻译科技文体时保持客观,做到不添加个人感情色彩,不歪曲、篡改原文内容。在科技翻译中,由于英语、汉语两种语言表达习惯不同,译者时常会采取增译或省译的方法处理原文。需要注意的是,为使译文语言更加符合目的语读者的阅读习惯,译者可以使用相关技巧对原文进行翻译,但是不能随意引申、删改,不能擅自增减原文的信息,以免译文与原文之间出现"差之毫厘,谬以千里"的现象。

(1) Last year, users expected to buy sparkling new 56K bit/sec. modems in hope of doubling their Internet access speeds.

译文一 去年,用户们期待购买新的亮丽的56K位/秒的调制解调器,希望能使他们访问因特网的速度提高两倍。

译文二 去年,用户们期待购买新的亮丽的56K位/秒的调制解调器,希望能使他们访问因特网的速度提高一倍。

分析 译文一将原文中的"doubling"处理为"提高两倍",属于误译。这一看似细小的错误会导致译文之意完全背离原文,让读者对调制解调器的能力的认识产生巨大偏差。英语中的"double""treble"等词作动词使用时指"使……变成两倍""使……变成三倍",即其表示的倍数都不包括基数在内。因此,原文中"doubling"的正确译法应是译文二中的"提高一倍"。由此可见,译者只有具备专业的知识,方能在正确理解原文的基础上译出忠实于原文的译文。

(2) Using data from ships, planes and satellites to study Asia's haze during the northern winter months of 1995 to 2000, scientists discovered not only that the smog cut sunlight, heating the atmosphere, but also that it created acid rain, a serious threat to crops and trees, as well as contaminating oceans and hurting agriculture.

译文一 科学家们通过飞船、飞机和卫星发回的1995年到2000年亚洲北部冬季期间的有关气象数据来对亚洲脏雾进行了分析研究。研究发现,这片烟雾不仅能阻断阳光射入和引起大气层升温,而且还能够引发酸雨,对农作物和树木构成威胁,还对海洋造成污染,对农业造成伤害。

译文二 科学家们通过飞船、飞机和卫星发回的1995年到2000年亚洲北部冬季期间的有关气象数据来对亚洲的雾霾进行了分析研究。研究发现,这里的雾霾不仅能阻断阳光射入和引起大气层升温,而且还能够引发酸雨,对农作物和树木构成威胁,还对海洋造成污染,对农业造成危害。

分析 译文一将“haze”与“smog”分别译为“脏雾”和“烟雾”,模糊、曲解了原文的意义,致使译文表意不明。首先,“脏雾”不仅有悖于汉语的表达习惯,而且其为何物也有待确认。与之相比,译文二将“haze”译为“雾霾”,清晰准确地还原了原文的意思。其次,译文一还将“smog”译为“烟雾”,同样有歪曲原文之嫌。译文二将两处“smog”均译为“雾霾”,既符合原文意思,也便于读者理解。由此可见,译者在翻译时应结合上下文语境,根据目的语的语言表达习惯,尽可能忠实地再现或反映原文之意。

第二节 通　顺

在忠实于原文的基础上,科技翻译还需要遵循通顺的原则与标准。这里的通顺指的是译文需具有逻辑性,符合目的语的语言表达习惯,读起来自然、流畅。为避免译文生硬、拗口,译者在忠实于原文的基础上可适当地调整用词、语序等,使其更符合目的语读者阅读习惯。需要注意的是,当通顺与忠实无法兼得时,可舍前者而取后者。

(1) Mineral products that were imported in large quantities included crude oil, iron ore, manganese ore, fine copper ore and potash fertilizer.

译文一 矿产大量进口,包括原油、铁矿、锰矿、细铜矿和钾肥。

译文二 原油、铁矿、锰矿、细铜矿和钾肥等矿产大量进口。

分析 两个译文均反映出原文之意,其中前者的语序与原文“高度一致”,但鉴于英汉两种语言表达习惯有别,相比之下,后者读起来更加自然、流畅。

(2) Mercury is a mental, although it is in the liquid state.

译文一 汞是金属,虽然它是液态的。

译文二 汞虽是液态,但在金属之列。

分析 译文一的语言结构与原文相同,但不太符合汉语的表达习惯。因为英语表让步关系时仅出现“although”或“but”,而汉语在相同情况下则通常使用“虽(然)/尽管……但(是)……”这一搭配。

第三节 准　确

鉴于科技文体具有精确性和细节性特征,准确是科技翻译中非常重要的原则与标准之一。科技文体中有许多专业词、半专业词和缩略语,这些词语翻译的准确性决定了译文的正确与否。在科技翻译中，一个词的误译可能会导致整个句子，甚至是整个语篇的意义与原文相背离。科技翻译中的错误，哪怕是再细微的差池，都可能会对科学研究和生产发展带来一定的负面影响。

(1)空气里弥漫着烟雾和尘埃。

译文一　The air is loaded with smoke and dirt.

译文二　The air is loaded with smog and dirt.

分析　两个译文分别将原文中的“烟雾”一词译为“smoke”和“smog”。《牛津英语词典》中将“smoke”解释为“the grey, white or black gas that is produced by something burning”(燃烧产生的灰色、白色或黑色气体)，将“smog”解释为“a form of air pollution that is or looks like a mixture of smoke and fog, especially in cities”(一种空气污染形式，是或像烟和雾的混合物，尤指在城市中存在)。由此可以看出“smoke”的含义更接近于汉语中的“烟”，而“smog”则更多指向汉语中的“烟雾”。相比之下，译文二更加达意。

(2)The bulk of the world's languages use base-10, base-20 or base-5 number systems.

译文一　世界上大部分语言使用以10、20或5为基础的数字系统。

译文二　世上的语言多采用以10、20或5为基数的数字系统。

分析　原文中的“base”属于第一章中提及的半专业词，即由普通词语通过意义扩展而形成的具有专业意义的词语。“base”一词在不同学科领域中词义不同，如在数学中指“基数”，在语言学中指“词根”，在棒球运动中指“垒”，在化学中指“碱”等。译文一按照“base”的基本词义将“base”译为“基础”，并不符合原文的意义。译文二根据“base”一词所在的上下文语境，将其译为“基数”，准确对应原文的意义。

(3)Streams and waterfalls are suitable for the development of hydroelectric power.

译文一　溪流和瀑布适合用于开发水电力量。

译文二　溪流和瀑布适于水电能源的开发。

分析　与例(2)类似，例(3)原文中的“power”一词也为半专业词，在不同的学科领域中词义不同(见第一章第一节之“二、单义性”)。译文一将其处理为“力量”，明显不符合原文之意，读起来也不太顺畅。译文二根据语境将其译为“能源”，准确再现原文之意。

第四节　简　洁

科技文体虽注重其细节性，但也绝不会忽视简洁性。科技文体中的语言表达简洁凝练，追求用最精练的语句表达尽可能多的信息。因此,简洁也是科技翻译的原则与标准之一。在翻译科技文体时,译者要在忠实于原文的基础上尽可能地精简语言,将原文的意义简明扼要地复现在译文中。

(1) The volume of the sun is about 1,300,000 times that of the earth.

译文一 太阳的体积约为地球的体积的1300000倍。

译文二 太阳的体积约为地球的130万倍。

分析 两个译文均准确传达出原文之意,但相比于译文一,译文二不仅删除了与前文重复的内容(“体积”),还将数字“1,300,000”凝练为“130万”,给人一目了然之感,此举更符合科技文体的要求。

(2) The emerging of uncertainties of input quantities leads to two different problems or questions.

译文一 输入量不确定性的出现导致两个不同的困难或问题。

译文二 输入量不确定性的出现导致两个不同的问题。

分析 “problems”和“questions”两词之意相近,译文一将二者均翻译出来,略嫌画蛇添足,且读起来不太符合汉语表达习惯,译文二则将其“合二为一”,简洁且清晰。

练习题

一、英译汉

1. The molecules combine, forming what is called a Schiff base within the protein.

2. He conjectured that the population might double in ten years.

3. There's a clear road map that speech recognition is going to move from niche to pervasive marketplaces.

4. When atmospheric pressure decreases, the boiling point becomes lower.

5. Although the per-unit cost of an SVC is higher than PVCs, if it is used less, the actual cost will be lower.

6. This heater needs a new element.

7. To examine stress distribution at the keyway and fracture surface, finite element method could be applied.

8. In order to make it safe, the element is electrically insulated.

9. Because sodium can increase blood pressure levels, lowering your consumption of salt and sodium may also contribute to better heart and vascular health.

10. With the development of the power industry, DEH governing system has been used increasingly.

二、汉译英

1. 黑色工具材料以铁为基材,包括工具钢、合金钢、碳钢、铸铁。

2. 大家都知道当今世界对石油的依赖到了何种程度。

3. 即使我们将盐溶于水也不能改变其化学性质。

4. 我们现在有 200 多万台机床,石油年产量超过 1 亿吨,煤炭超过 6 亿吨,钢材仅有 3000 多万吨。

5. 由于摩擦作用,这种橡胶密封圈容易损坏。

6. 地壳系统中,地震发生的位置被称作震源。

7. 金刚钻是已知可以用作切割刀具的最硬材料。

8. 虚拟技术广泛应用于各种汽车试验平台。

9. 车身的振动通过悬架和轮胎传递到地面。

10. 电动汽车领域最新的技术进步主要集中在能源效率的提高上。

第三章
译者素养

从事翻译,译者需具备多方面素养。作为科技文章的译者,亦是如此。林语堂在《论翻译》一文中认为译者应"对于原文文字及内容上(有)透彻的了解;有相当的国文程度,能写清顺畅达的中文;对于翻译标准及手术的问题有正当的见解"。王佐良在《翻译中的文化比较》一文中提出:"翻译者必须是一个真正意义的文化人。首先他必须掌握两种语言,但不了解语言当中的社会文化,他就无法真正掌握语言。他不仅需要深入了解外国文化,而且还要深入了解自己民族的文化,更重要的是他还要不断地把两种文化加以比较。因此在翻译中,他处理的是个别的词,他面对的则是两大片文化。"由此我们可以看出,在翻译中,译者需要通晓源语,熟练运用目的语,熟悉源语与目的语背后的文化,了解翻译的基本知识与操作。

除了具备译者需具备的基本素养外,科技翻译中译者还应具备哪些素养呢?晚清时期,国外科技译者傅兰雅通过西学翻译将诸多西方近代科学技术知识引入中国。根据傅兰雅的相关言论,可以将他对科技翻译译者素养的要求归纳为三点,分别是对所译学科知识的沉淀、对语言文化的了解以及对翻译准确性的追求。由此可见,科技翻译因文体的特殊性而对译者有了特殊的要求。我们可以将科技翻译中译者所需具备的素养归纳总结为四点:语言素养、职业素养、专业素养、工具素养。

第一节 语言素养

科技翻译工作者首先需要具备一定的语言素养。这里的语言素养指科技翻译工作者应通晓源语与目的语及其背后分别对应的文化。本教程主要讨论英汉科技翻译,在翻译时,译者应熟练掌握英语与汉语两种语言并了解语言背后的文化。

首先,科技翻译工作者需具备扎实的英语基础,以及听说读写的能力。由于科技英语多使用长句,译者在翻译科技文体时需具备良好的英语阅读能力,以便很好地理解和分析原文中的长句。面对长句的翻译,译者需先厘清长句中各个分句间的逻辑关系,如哪些是并列关系、哪些是修饰限定关系、哪些是因果关系、哪些是让步关系等。在明确长句中各成分之间的语法、语义关系后再进行翻译,以避免误译,提高翻译的准确率。另外,科技翻译工作者还需具备良好的英语写作能力，在汉译英的过程中能够保证译文的准确、流畅并符合英语本族语者的阅读习惯。

(1) Apple is providing a free Macintosh application, dubbed iBooks Author, which allows publishers, teachers and writers to produce interactive textbooks with video, audio and even rotating 3D graphics that spring to life with the touch of a finger.

译文一 苹果公司提供一个免费的苹果电脑应用，被称为电子书创作程序，这种程序允许出版商、教师和作家创作视频、音频，甚至旋转3D图形的交互式课本，这些课本用手指就能操控。

译文二 苹果公司正在提供一款名为iBooks Author的免费苹果电脑应用程序，出版商、教师和作家可以用它制作带有视频、音频甚至旋转3D图形的交互式课本，只要用手指一触，这些3D图形就会动起来。

分析 在翻译该句时，译者只有厘清这一长句中各个分句之间的关系以及介词短语间的关系，才能确保译文与原文逻辑关系一致。首先是“which”用作关系代词引导的非限制性定语从句修饰限定“iBooks Author”这一应用程序，其次是“with”引导的介词短语修饰限定“interactive textbooks”，最后是“that”用作代词引导限制性定语从句，修饰限定“rotating 3D graphics”。分析清楚原文中各成分之间的关系后，我们可以看出译文一带有明显的“翻译腔”，且最后“that”从句的汉译有误。相比之下，译文二则更加准确、自然。

(2)这就提出了一个非常有意思而又密切相关的问题：一些针对免疫系统的药物可能对急性髓细胞也有影响。

译文 That raises the very interesting and pertinent question that some drugs that target the immune system might have an effect on AML cells.

分析 在翻译该句时，译者在理解原文意义的情况下应思考清楚如何翻译才能使译文自然、流畅，符合英语本族语者的阅读习惯。由于科技英语中常使用长句，将该句翻译为一个逻辑清晰、关系清楚的长句是译者的较佳选择。译文中使用了三个“that”将原文串联起来。其中，第一个“that”为代词，用作主语，将原文中的“这”还原出来，指代上文中提到的某一信息；第二个“that”为连词，引导同位语从句，解释说明“question”的内容，充当原文中冒号的功能；第三个“that”为代词，引导限制性定语从句，由于科技英语中常出现定语后置现象，此处将原文中“一些针对免疫系统的药物”处理为“that”引导的后置定语从句，修饰并限定前面的“some drugs”，使其更加符合英语表达习惯。

其次，科技翻译工作者需具备相当的汉语表达水平。在科技翻译中，译者既需要保证忠实于原文，不随意歪曲、增删原文，还需要保证译文的通顺、流畅、自然。科技翻译工作者在英译汉的过程中，需有较高的汉语水平，从而避免“翻译腔”的出现，使译文更加符合汉语的表达习惯。

(3) Scientists have directly measured the moisture in the air and confirmed that it is rising, supplying the fuel for heavier rains, snowfalls and other types of storms.

译文一 科学家们直接测量了空气中的湿度,并证实它正在上升,为更强的降雨、降雪和其他类型的风暴提供了燃料。

译文二 科学家们直接测量了大气中的湿度,并证实其正在上升,这一现象正是更大规模降雨、降雪和其他风暴类型的源头。

分析 译文一将“supplying the fuel”按照字面意思直接译为“提供了燃料”,虽符合原文意义,但整个译文的措辞有悖于汉语表达习惯。相较而言,译文二将其译为“是……的源头”,不仅符合原文意义,也符合汉语表达习惯,读起来更加通顺、流畅、自然。

第二节 职业素养

作为一名科技翻译工作者,无疑应具备普通译者的职业素养,即熟悉翻译理论,了解翻译原则、标准与目的,掌握翻译策略与翻译技巧并保持严谨专业的工作态度。专业的翻译理论知识与实践技能是区分职业译者与非职业译者的关键。熟悉 Nida 的功能对等理论、Reiss 和 Nord 的功能翻译理论、Vermeer 的目的论、Newmark 的语义翻译和交际翻译理论以及严复的“信达雅”、钱钟书的“化境说”、傅雷的“神似观”等,同时练就增译、省译、拆译、合译、词性转换、反译等翻译方法,对于清晰、准确、流畅地进行翻译起到不可替代的指导作用。

(1) When voltage is applied to a metallic layer between the plates, it warms up and radiates heat to the room.

译文 当电流通过玻璃板之间的金属层,加热器升温并将热量向房间辐射。

分析 译者通过采取增译法,将原文中的代词“it”翻译为看起来更加完整、明确的名词“加热器”,使其更加具体、指向明确。由此可以看出,在忠实于原文意义的情况下,采用适当的翻译方法可以使得译文更加清晰、自然。

此外,鉴于科技文体精确性的特征,科技翻译工作者在翻译时应持有专业态度,认真严谨对待翻译工作。在进行科技翻译时,译者需严肃认真,做到字字计较、词词斟酌,避免出现低级错误和疏漏以及任何对原文的歪曲与篡改。因为在科技翻译中,任何细微的错误都有可能对知识传播与生产生活造成影响和损失。

(2) This is the full differential equation for the full mechanism. Not just one part of it.

译文一 这是整个机理的完整的不同方程,不是它的一部分。

译文二 这是整个机理的完整的微分方程，而非一部分。

分析 译文一的译者显然是未仔细推敲原文的意义，未根据原文语境分析其中“differential”的意义，而是将其想当然地理解为“不同的、有差别的”。这样的错误看起来很微小，但完全改变了原文意义，使译文读者无法准确地理解原文想要表达、传递的信息。引发该错误的原因在于译者对科技翻译工作不够用心，态度不够端正，准备不够充分。但凡译文一的译者拥有一定的知识储备，知晓“differential equation”指“微分方程”，或是对“不同方程”一说持有些许疑问，进行查询与检索，就不至于出现类似错误。

第三节 专业素养

科技文体主要指谈及科学和技术的书面语和口语，其内容主要涉及科学、技术、工程等专业领域，具有专业性、客观性、精确性、细节性和逻辑性五大特征。科技文体所涉领域广泛，要求译者具备广博的知识面，对各个领域都有一定的了解，掌握相关专业的基本知识，尤其是相关领域的基本词汇和常用词汇。比如，在翻译有关物理知识之时，需要知道“加速度”是“acceleration”，“交流电”是“alternating current”，“阳极、正极”是“anode”，“衰减”是“attenuation”，“平均功率”是“average power”，“浮力”是“buoyancy”，等等；在翻译有关数学内容时，需要知道“集合”是“aggregate”，“代数”是“algebra”，“公因数”是“common factor”，“常数”是“constant”，“小数”是“decimal”，“分母”是“denominator”，“函数”是“function”，“积分”是“integral”，等等。译者应了解与科技翻译内容相关的专业知识，拓宽知识面，完善知识结构，从而有效提高翻译质量与效率。此外，由于科技文体中存在较多专业词、半专业词和缩略词，科技翻译工作者还需有意识地结合上下文语境去辨别相关词义。特别是对于在不同学科领域具备不同意义的半专业词汇以及缩写形式相同的缩略词，更需仔细辨别其在文内所指代的意义，从而避免不必要的误译。

（1）镰状细胞贫血通过隐性基因遗传给后代。

译文 Sickle-cell anaemia is passed on through a recessive gene.

分析 在翻译原文时，如果译者具备相应的医学、生物学知识，同时知晓“镰状细胞贫血”和“隐性基因”的英文对应表述，那么翻译这句话便是信手拈来。倘若译者没有相关的知识储备，对这些术语的对应翻译不甚了解，这时最为稳妥的方式就是求助相关专家或认真查阅词典。

（2）This is a single cell of a battery.

译文 这是一个单芯的电池。

分析 “cell”一词最为常用的意义是“细胞”，如果译者没有相关的知识储备，不了解

"cell"在不同学科中的不同意义,则极有可能造成误译。与之相反,如果译者知晓"cell"一词在不同学科中的不同意义,结合上下文语境,便能轻松判别其在这里指"电芯"了。

众所周知,科学技术是第一生产力。当今社会,各个国家都在不断加大对科学技术的投入,鼓励科学技术创新。在科学技术不断发展的大背景下,科技翻译的具体对象实时更新,因此,科技翻译工作者还需要与时俱进,不断更新相关知识储备,跟上科技发展的步伐。

第四节　工具素养

鉴于科技文体的专业性以及科技翻译对于准确性的要求,科技翻译工作者在不断增强自身专业知识储备的同时,还需要具备一定的工具素养,学会利用相关工具来辅助翻译工作。由于译者专业知识的有限性与科技文体领域的广泛性以及内容的多样性之间存在矛盾,译者会不可避免地遇到难以处理的字词。在这种情况下,译者应充分利用现有工具,提高翻译的质量和效率。

首先,科技翻译工作者需要善用工具书,如字典词典、百科全书、手册、标准等。其中,字典词典的使用尤为重要,特别是翻译术语等专业词、半专业词和缩略词等时。字典词典能够弥补译者专业词汇量的不足,帮助译者确定词语的正确意义,提高翻译的准确率。

其次,科技翻译工作者还需熟练运用网络,利用网络了解原文所涉及专业领域的最新发展以及相关基本知识。网络可以快速、便捷地为科技翻译工作者提供其所需的材料内容。针对一些最新的科技知识,可能相关字典词典还未更新专业词条,但有时网络上已经出现了这些新术语的相关翻译,可以为译者提供参考。特别是 ChatGPT 的出现进一步更新了人工智能和搜索引擎的发展,能够方便译者更加快捷地找到自己所需的答案。

最后,科技翻译工作者还需掌握机计算机辅助翻译(Computer-Aided Translation,CAT)工具。随着科学技术的不断发展,人类的生产生活中充满了科技的身影,各行各业的工作人员都能够利用相关科技成果提高工作效率,科技翻译同样如此。计算机辅助翻译狭义上指为翻译任务及其管理专门设计的计算机工具,以翻译记忆技术为核心,用以提高翻译效率、优化翻译流程。与机器翻译不同,计算机辅助翻译的翻译过程仍以人为主导,主要功能有句段切分、翻译记忆库、术语库、质量保证等。目前较为常见的计算机辅助翻译工具有 SDL Trados(https://www.sdltrados.cn/cn/)、Déjà Vu(https://atril.com/)、memoQ(https://www.memoq.com/)、OmegaT(https://omegat.org/zh_CN/)、Wordfast(https://www.wordfast.com/)、MateCat(https://www.matecat.com/)、Memsource(https://www.memsource.com/)、Smartcat(https://www.smartcat.com/)、Transmate(http://www.uedrive.com/products/

standalone/)、快译点(http://www.91kyd.com/index.html)、雪人CAT(http://www.gcys.cn/index.html)、YiCAT(https://www.yicat.vip/)、译马网(https://www.jeemaa.com/#/portalPage/home)、云译客(https://pe-x.iol8.com/home)等。其中,以SDL Trados最为流行,使用范围广泛。计算机辅助翻译工具可以减轻科技翻译工作者的工作负担,帮助提高翻译效率,减少译者的重复劳动。在许多专业术语和统一句式的处理上,计算机辅助翻译工具中的翻译记忆库、术语库等功能能帮助译者节省大量时间。在其辅助之下,译者可以将更多时间用于修改润色译文,从而提高科技翻译的效率和质量。

练习题

一、英译汉

1. Unlike computers, which are installed in classrooms and shared by pupils, the whole point of a tablet is that it is carried around by an individual and used anywhere, including the home.

2. Instead of exhaust fumes, the fuel-cell car produces soap and water.

3. The DNA molecule is compounded from many smaller molecules.

4. The USDA Food Patterns and DASH Eating Plan are healthy eating patterns that provide flexible templates, allowing all Americans to stay within calorie limits, meet their nutrient needs, and reduce chronic disease risk.

5. An analog signal is composed of four basic components: DC and AC magnitudes, frequency, and phase.

6. Single crystals of high perfection are an absolute necessity for the fabrication of integrated circuits.

7. HEVs have the potential to reduce fuel consumption and emissions in comparison to conventional vehicles.

8. Noise means any unwanted sound.

9. Whether iron comes in contact with water, or with moisture, both make it rusty.

10. The temperature in the furnace is not always above 1,000℃.

二、汉译英

1. 药剂师使用什么剂量单位?

2. 并且在这个充满火箭弹、反火箭弹拦截导弹、雷达、控制室、无人机和无人机黑客的日趋高科技的战场,正是像Idan Yahya(不论他在阿拉伯国家的对手是谁)这样的士兵发挥着最大的作用。

3. 温度不变,所有分子的平均速度就不变。

4. 直到喷气发动机发明后,飞机才能以超音速的速度飞行。

5. 在混凝土中起化学作用的是水泥而不是集料。

6. 实现这种改变最简单的方式是换用柴油机。

7. 不但晶体管坏了,线路也出了故障。

8. 在选择电气材料时,电导率十分重要。

9. 科技产业之外的领域也因芯片短缺而受到影响。

10. 根据“烟囱效应”原理,空气通道密度越高,热压通风量越大。

第四章

网络查询工具与方法

当今科技日新月异，科技翻译涉及的专业名词、动词等词语也极为浩瀚。虽然谷歌翻译、百度翻译、必应翻译、DeepL 等各类机器翻译平台及 ChatGPT、“我在 AI”等人工智能平台能帮助译者大幅节省初译时间，但机器翻译或人工智能翻译版本中的欧化中文句式及中式英语现象仍然比较明显，术语译文也存在不少问题。在此背景下，如何高效利用电子词典、普通在线网页、语料库等资源查询、验证科技词汇显得至关重要。科技翻译界普遍认同的观点是：好翻译是查出来的，要想成为翻译高手，首先得成为查询高手。

第一节 在线词典

灵格斯词霸

灵格斯(Lingoes)词霸是一款简明易用的词典与文本翻译软件，支持全球 80 多种语言的词典查询、全文翻译、屏幕取词、划词翻译、例句搜索、网络释义和真人语音朗读功能。同时还提供海量词库免费下载，专业词典、百科全书、例句搜索和网络释义一应俱全。软件官网(http://www.lingoes.cn/zh/dictionary/)提供各类语言词典和专业词典下载服务，但美中不足的是，网站提供的桌面软件只适用于 Windows 系统。

灵格斯词霸支持安装多个词典，译者可将常用的语言词典和专业领域词典一起安装，这样就能在一个界面显示多个词典的解释。安装后，可在词典管理处设置索引组和取词组的词典，并设置搜索结果先后顺序。以《英汉航海词典》为例，安装后在“词典管理”中选择该词典，点击按钮，将其置顶(图 4-1 中箭头所示)，以便快速找到该领域词语的翻译。

以航海领域文本翻译为例，在主界面输入“lighter”或选中翻译文本中的该词(划词翻译)，即可找到航海领域的唯一意思“驳船”，而其他非专业词典的释义涉及领域会相当广泛(图 4-2)，可能会使译者产生困惑。

图 4-1 灵格斯词霸安装界面

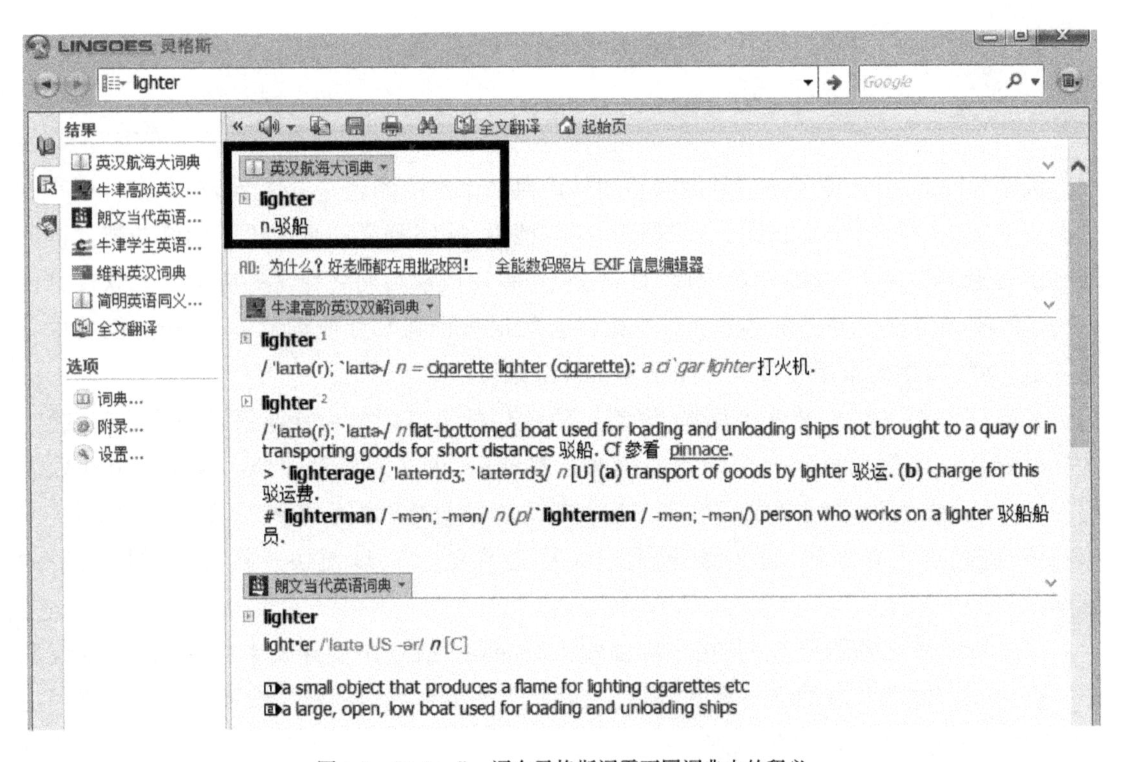

图 4-2 “lighter”一词在灵格斯词霸不同词典中的释义

二 欧路词典

欧路词典支持学习笔记、生词本多平台同步,其主要功能和灵格斯词霸大同小异,但可以导入和查询牛津高阶第八版等年代稍近一些的词典。该词典提供 Mac、Windows、Linux 三种电脑系统的桌面安装版及苹果(iPhone、iPad)、安卓(Android)两种系统的手机版(图 4-3)。欧路词典的主要使用方法和灵格斯词霸相近,在此不做详细介绍。

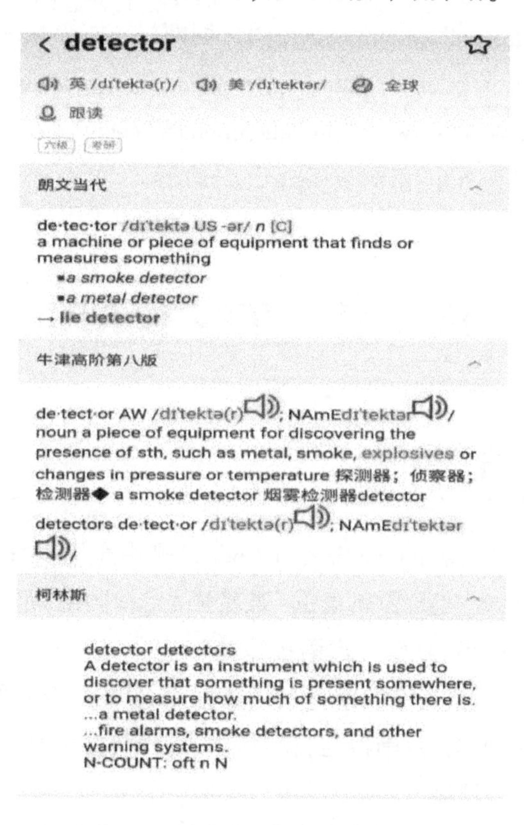

图 4-3 欧路词典安卓手机版查词演示

三 MDict 词典

MDict 词典(https://www.mdict.cn/)是一款开放式词典读取软件,可支持多平台使用,用户可以自行制作词典,还可以实现简繁体汉字转换。该软件界面简洁,操作方便易懂。用户可通过 pdawiki 论坛(https://www.pdawiki.com/forum/)搜索自己需要的词典并安装至 MDict 词典软件中。

对译者而言,MDict 词典的主要功能与灵格斯词霸、欧路词典等软件相似,可以导入语言类词典和专业词典。

四 网页类词典

常见的网页类词典有《朗文当代英语辞典》(在线)(Longman Dictionary of Contemporary English Online, https://www.ldoceonline.com)、《牛津学生英语辞典》(Oxford Learner's Dictionaries, https://www.oxfordlearnersdictionaries.com)、《柯林斯辞典》(Collins Dictionary, https://www.collinsdictionary.com)、《牛津在线搭配词典》(Online Oxford Collocation Dictionary, https://www.freecollocation.com)、同义词网(https://www.thesaurus.com)、《韦氏大词典》(Merriam-Webster Dictionary, https://www.merriam-webster.com)、《麦克米伦词典》(Macmillan Dictionary, https://www.macmillandictionary.com)、《汉语词典》(https://cidian.bmcx.com)、《汉语大辞典》(http://www.hydcd.com)等。

第二节　在线术语库

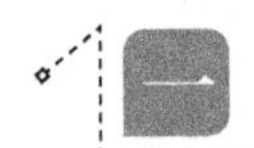

术语在线

"术语在线"(https://www.termonline.cn/index)是由全国科学技术名词审定委员会主办的规范术语知识服务平台,是规范术语的"数据中心""应用中心"和"服务中心",提供术语检索、术语管理(纠错、征集、分享)、术语提取与标注、术语校对等服务。术语在线包含了全国科学技术名词审定委员会发布的规范名词数据库、名词对照数据库、工具书数据库等资源,累计有50万余条规范术语,范围覆盖自然科学、工程与技术科学、医学与生命科学、人文社会科学、军事科学等学科领域。

术语在线平台操作简便,在首页输入想要翻译的术语的所有字段或重要字段,即可查询结果。如输入"表面机械强化"(图4-4),能清楚地得知该术语为2011年公布的材料科学技术名词,其对应的公认英文术语为"mechanical surface strengthening"。需要注意的是,"表面机械强化"在DeepL翻译、谷歌翻译、ChatGPT、百度翻译中的译文为"surface mechanical strengthening",只有在有道翻译中的译文为"mechanical surface strengthening"。这说明,机器翻译平台处理专业术语的能力仍然不够专业、不够强大,需要人工多方验证。

再如,在翻译生物工程技术相关文本时遇到"基因敲除"这个术语,可尝试在术语在线中输入该词,可得到"gene knockout""gene knock-out""gene knock out"三种说法(图4-5a)。

将上述页面下拉还会发现,"基因敲除"又称"基因剔除"(图4-5b)。如果今后遇到"gene knockout",就不用为选择"基因敲除"还是选择"基因剔除"而费神了。译者往往是在翻译过程中顺便学习新知识，为今后提高翻译效率储备知识。

图 4-4 “表面机械强化”在术语在线中的译名截图

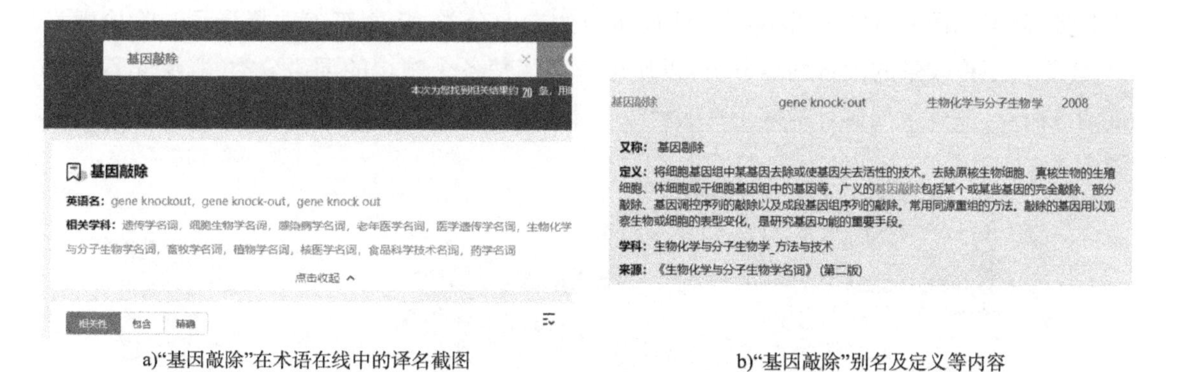

a)“基因敲除”在术语在线中的译名截图

b)“基因敲除”别名及定义等内容

图 4-5 “基因敲除”在术语在线中的查询结果

值得注意的是，术语在线中收录的多为名词短语和偏正短语，动宾短语很少，单一动词也极少，要确定某一个动词往往还需要借助其他工具。

二 中国规范术语库

中国规范术语库——知网学术百科（http://shuyu.cnki.net）是中国知网和全国科学技术名词审定委员会的合作项目，旨在帮助专业工作者规范、正确使用本领域的专业术语，提高专业水平。该网站首页中的术语库包括自然科学、农业科学、医药科学、工程与技术科学、人文与社会科学五大板块,每个大板块下也有细化的小分类。因该网站的术语主要是根据全国科学技术名词审定委员会历年审定公布并出版的数据制作,其操作与“术语在线”无明显差别,故此处不再举例介绍。

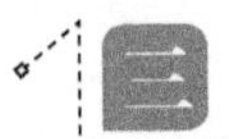

三 WIPO Pearl 术语库

WIPO 是世界知识产权组织的英文缩写,Pearl 形容其内容十分珍贵。WIPO Pearl 术语库(https://wipopearl.wipo.int/en/linguistic)拥有强大的专利数据库,在查找科技术语等方面具有优势,可保证某术语译本规范统一。该工具提供的不只是所查询术语语言上的转换服务,还提供了对术语的解释及每个术语出现的专业领域。概念图有助于译者在查询时更好地理解该术语,同时译者对于不完全确定的术语还可以尝试模糊搜索。其缺点是仅收录了专利文件中的科技词汇,因此,对于未申请专利的技术及非技术类概念词汇收录有限。

四 SCIdict 学术词典

SCIdict 学术词典(http://www.scidict.org/)主要面向研究人员、各行业专业人士以及专业翻译人员。通过 SCIdict,可准确查询各专业领域的术语,并能标记每个术语词条所属的词语分类,使得翻译人员能够准确判断译词所属的专业分类,从而正确选择译词。除检索术语,该平台还将收录的大量词语分类,以便用户更清楚各个领域的词语分类,避免学科词语混淆。因此,该学术词典可谓是个大型科技学术术语库。例如,在该网站输入“页岩”,能得到图 4-6 所示结果。

页岩

[数理] 含矿页岩 ore bearing shale 收藏

图 4-6 “页岩”一词在 SCIdict 学术词典中的搜索结果

再以“heterozygous mutation”为例查询其汉译名。从图 4-7 所示结果可以看出,

“heterozygous mutation”可以译为“杂合子突变”,也可以译为“杂合突变”。

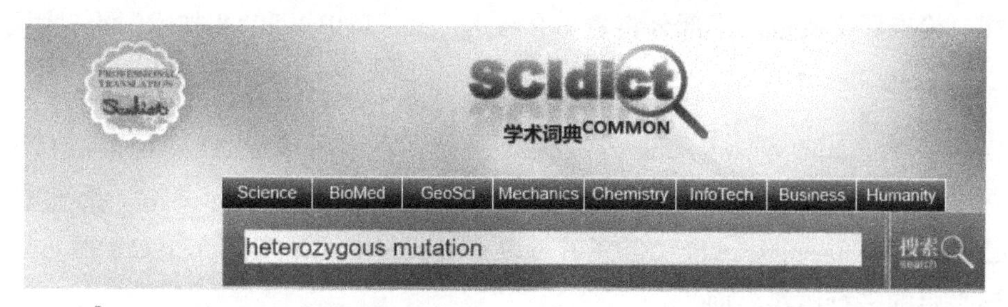

heterozygous mutation

[生物医学] heterozygous mutation 杂合子突变 收藏 ➡

[生物医学] compound heterozygous mutation 复合杂合突变; 杂合 收藏 ➡

[生物医学] Heterozygous 缺失; 杂合子; 异质结合的; 畸杂合的; 异形接合体; 杂合的 收藏 ➡

[生物医学] Heterozygous hybrid 远缘杂交 收藏 ➡

[生活] heterozygous gene 杂合子 收藏 ➡

[生物医学] heterozygous sex 性染色体异型的性别 收藏 ➡

[生物医学] heterozygous bacterial cells 异型接合细菌细胞 收藏 ➡

[生物医学] compound heterozygous mutation 复合杂合突变 收藏 ➡

图 4-7 “heterozygous mutation”一词在 SCIdict 学术词典中的搜索结果

五 中国大百科全书

中国大百科全书(第三版,网络版,https://www.zgbk.com/)于 2023 年集中发布,共发布 50 万个网络版条目、10 卷纸质版图书,修订更新了部分中国主题英文条目。该网站分为专业版、专题版和大众版 3 个板块,涵盖国家颁布的所有知识门类和一级学科,按照 94 个执行学科和近百个专题进行编纂。随文配置图片、公式表格、视频、音频、动画,并附有知识链接,内容丰富、形式多样,便于读者浏览和学习。

第三节 语 料 库

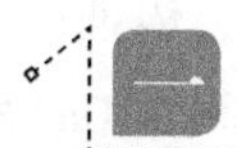

一 双语语料库

双语语料库检索便利、储存量大,在译员作品风格、机器翻译等方面发挥着极为重要的作用。双语语料库拥有大量原文与译文实例，能够在更大范围内方便译员或校对人员查找语言搭配、检查译文质量。本部分将着重介绍 Linguee、LetPub 和“X 技术”网三大网站。

（一）Linguee

Linguee 平台（https://www.linguee.com/）支持多种语种查词，能满足英语和小语种使用者群体的需求。该平台的词语翻译功能还提供大量双语平行文本。虽然页面显示“External Sources (not reviewed)”，但译者可以查看其权威网站来源。图 4-8 展示了“钢筋”一词在该平台的平行语料界面。

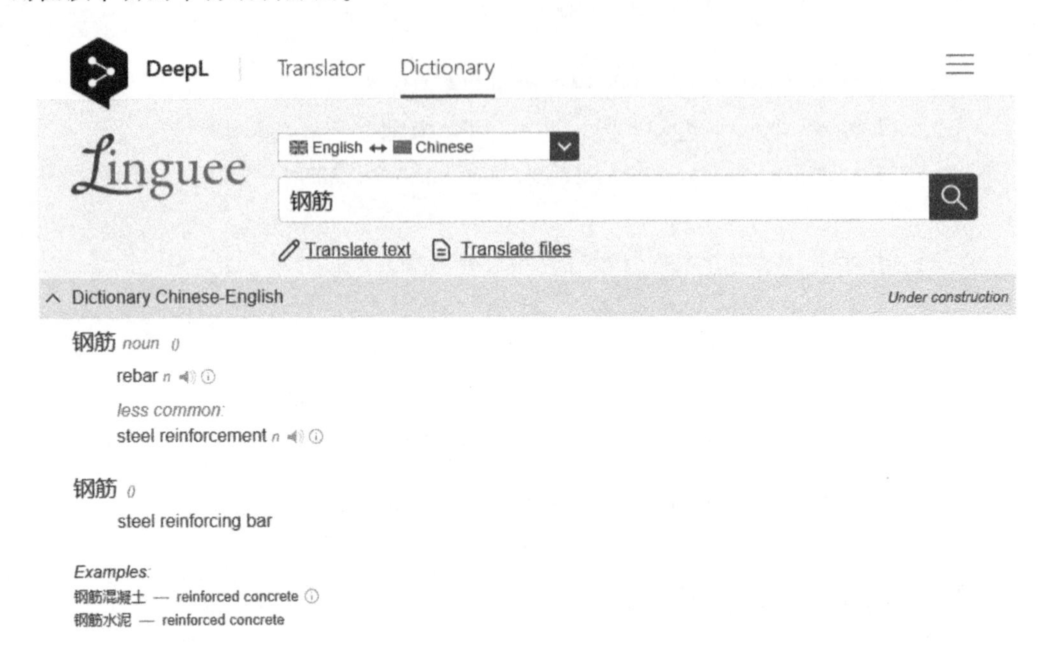

图 4-8　Linguee 网站中“钢筋”的搜索结果

从图 4-8 可以看出，单独以“钢筋”一词出现时，一般英语用“rebar”，也可以用“reinforcement”，而“钢筋”在偏正结构中作形容词时，英语为“reinforced”。建筑领域翻译的译者能够判断这些是否准确，对该行业不熟悉的译者可以下滑页面，找到如图 4-9 所示的平行语料。其中，不乏有来自“联合国”（图中 daccess-ods.un.org）、“中国建筑师”（图中 chinese-architects.com）等权威网站的双语语料。

（二）LetPub

LetPub（http://www.letpub.com.cn/）是美国 ACCDON 公司旗下的专业品牌，为非英语国家科研学者提供优质 SCI 论文编辑和各类相关服务。LetPub 推出的在线翻译词典汇集了最全面的专业医学英语词汇翻译和 SCI 论文对照功能，包括中医常用医学名词、传染病学名词、儿科医学名词、内科学专业名词等。例如，在网站输入“步态”，可得到如图 4-10a）所示结果。

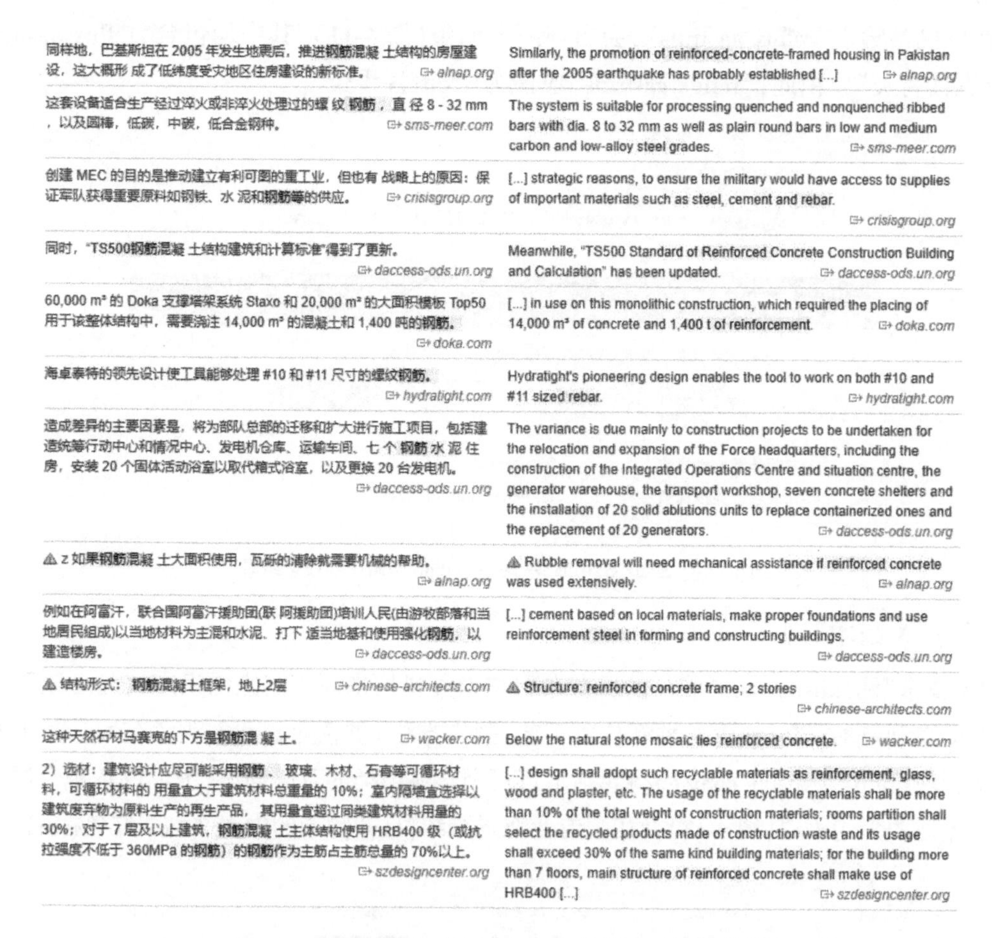

同样地，巴基斯坦在 2005 年发生地震后，推进钢筋混凝 土结构的房屋建设，这大概形 成了低纬度受灾地区住房建设的新标准。 alnap.org	Similarly, the promotion of reinforced-concrete-framed housing in Pakistan after the 2005 earthquake has probably established [...] alnap.org
这套设备适合生产经过淬火或非淬火处理过的螺 纹 钢筋，直 径 8 - 32 mm ，以及圆棒，低碳，中碳，低合金钢种。 sms-meer.com	The system is suitable for processing quenched and nonquenched ribbed bars with dia. 8 to 32 mm as well as plain round bars in low and medium carbon and low-alloy steel grades. sms-meer.com
创建 MEC 的目的是推动建立有利可图的重工业，但也有 战略上的原因：保证军队获得重要原料如钢铁、水 泥和钢筋等的供应。 crisisgroup.org	[...] strategic reasons, to ensure the military would have access to supplies of important materials such as steel, cement and rebar. crisisgroup.org
同时，"TS500钢筋混凝 土结构建筑和计算标准"得到了更新。 daccess-ods.un.org	Meanwhile, "TS500 Standard of Reinforced Concrete Construction Building and Calculation" has been updated. daccess-ods.un.org
60,000 m² 的 Doka 支撑塔架系统 Staxo 和 20,000 m² 的大面积模板 Top50 用于该整体结构中，需要浇注 14,000 m³ 的混凝土和 1,400 吨的钢筋。 doka.com	[...] in use on this monolithic construction, which required the placing of 14,000 m³ of concrete and 1,400 t of reinforcement. doka.com
海卓泰特的领先设计使工具能够处理 #10 和 #11 尺寸的螺纹钢筋。 hydratight.com	Hydratight's pioneering design enables the tool to work on both #10 and #11 sized rebar. hydratight.com
造成差异的主要因素是，将为部队总部的迁移和扩大进行施工项目，包括建造统筹行动中心和情况中心、发电机仓库、运输车间、七 个 钢筋 水 泥 住房，安装 20 个固体活动浴室以取代箱式浴室，以及更换 20 台发电机。 daccess-ods.un.org	The variance is due mainly to construction projects to be undertaken for the relocation and expansion of the Force headquarters, including the construction of the Integrated Operations Centre and situation centre, the generator warehouse, the transport workshop, seven concrete shelters and the installation of 20 solid ablutions units to replace containerized ones and the replacement of 20 generators. daccess-ods.un.org
⚠ z 如果钢筋混凝 土大面积使用，瓦砾的清除就需要机械的帮助。 alnap.org	⚠ Rubble removal will need mechanical assistance if reinforced concrete was used extensively. alnap.org
例如在阿富汗，联合国阿富汗援助团(联 阿援助团)培训人民(由游牧部落和当地居民组成)以当地材料为主混和水泥、打下 适当地基和使用强化钢筋，以建造楼房。 daccess-ods.un.org	[...] cement based on local materials, make proper foundations and use reinforcement steel in forming and constructing buildings. daccess-ods.un.org
⚠ 结构形式： 钢筋混凝土框架，地上2层 chinese-architects.com	⚠ Structure: reinforced concrete frame; 2 stories chinese-architects.com
这种天然石材马赛克的下方是钢筋混 凝 土。 wacker.com	Below the natural stone mosaic lies reinforced concrete. wacker.com
2）选材：建筑设计应尽可能采用钢筋 、玻璃、木材、石膏等可循环材料，可循环材料的 用量宜大于建筑材料总重量的 10%；室内隔墙宜选择以建筑废弃物为原料生产的再生产品， 其用量宜超过同类建筑材料用量的 30%；对于 7 层及以上建筑，钢筋混凝 土主体结构使用 HRB400 级（或抗拉强度不低于 360MPa 的钢筋）的钢筋作为主筋占主筋总量的 70%以上。 szdesigncenter.org	[...] design shall adopt such recyclable materials as reinforcement, glass, wood and plaster, etc. The usage of the recyclable materials shall be more than 10% of the total weight of construction materials; rooms partition shall select the recycled products made of construction waste and its usage shall exceed 30% of the same kind building materials; for the building more than 7 floors, main structure of reinforced concrete shall make use of HRB400 [...] szdesigncenter.org

图 4-9 搜索"钢筋"一词时在 Linguee 页面出现的双语语料

从图 4-10a)可知，医学领域的"步态"全部译为"gait"。此外，图 4-10a)中的中英文词条均可点击进入，从已发表的医学 SCI 原文(节选)中获取部分语料(图 4-10b)。

a)"步态"一词在LetPub中的搜索结果

b)"步态"英文"gait"在LetPub中的语料

图 4-10 "步态"一词在 LetPub 网站中的搜索结果

在该网站输入"reflux gastritis",会出现两个词条(图4-11),其中均包含"reflux gastritis"和"反流性胃炎"。由此,"reflux gastritis"应译为"反流性胃炎"。

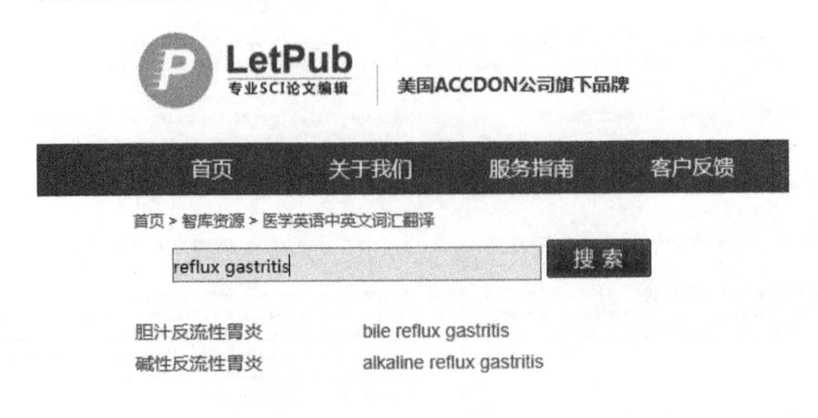

图4-11 "reflux gastritis"一词在LetPub网站中的搜索结果

(三)"X技术"网

"X技术"网(xjishu. com)收录的所有专利技术资料均来源于国家知识产权局。该网站提供最新的专利技术资料(包含各种新技术工艺、制作方法)供企业家和创业者查询。其中的科技英语板块(http://xjishu. com/en/)囊括了近百个学科门类的双语词汇(图4-12),每个词条里还有大量双语短句语料(图4-13)及相关专业背景介绍,且关键词的中英文均用红色标记。

科技英语				
天文学学术英语	外科学学术英语	微生物学学术英语	文化理论学术英语	文学理论学术英语
无线电电子学,电信技术学术英	物理学学术英语	戏剧艺术学术英语	细胞生物学学术英语	心理学学术英语
信息与知识传播学术英语	畜牧,动物医学,狩猎,蚕,蜂学术英	眼科学学术英语	药学学术英语	冶金工业学术英语
一般工业技术学术英语	医学研究方法学术英语	遗传学学术英语	艺术理论学术英语	音乐舞蹈学术英语
邮电经济学术英语	语言学学术英语	预防医学,卫生学学术英语	园艺学术英语	原子能技术学术英语
灾害及其防治学术英语	哲学理论学术英语	植物保护学术英语	植物学学术英语	中国文学学术英语
中国医学学术英语	中国哲学学术英语	肿瘤学学术英语	自动化技术,计算机技术学术英	自然地理学学术英语
自然科学学术英语	安全科学学术英语	财政,金融学术英语	测绘学学术英语	常用外国语学术英语
大气科学(气象学)学术英语	地球物理学学术英语	地质学学术英语	电工技术学术英语	电影,电视艺术学术英语
动物学学术英语	儿科学学术英语	耳鼻咽喉科学学术英语	法律学术英语	废物处理与综合利用学术英语
分子生物学学术英语	妇产科学学术英语	工业经济学术英语	工艺美术学术英语	古生物学学术英语
海洋学学术英语	汉语学术英语	航空,航天学术英语	化学学术英语	化学工业学术英语
环境保护管理学术英语	环境科学基础理论学术英语	环境污染及其防治学术英语	环境质量评价与环境监测学术英	绘画学术英语
机械,仪表工业学术英语	基础医学学术英语	建筑科学学术英语	交通运输学术英语	交通运输经济学术英语
教育学术英语	金属学与金属工艺学术英语	经济计划与管理学术英语	经济学学术英语	晶体学学术英语
科学,科学研究学术英语	口腔科学学术英语	矿业工程学术英语	昆虫学学术英语	力学学术英语
历史,地理学术英语	林业学术英语	临床医学学术英语	旅游经济学术英语	伦理学(道德哲学)学术英语
逻辑学(论理学)学术英语	贸易经济学术英语	美学学术英语	内科学学术英语	能源与动力工程学术英语
农学(农艺学)学术英语	农业工程学术英语	农业基础科学学术英语	农业经济学术英语	农作物学术英语
欧洲哲学学术英语	皮肤病学与性病学学术英语	普通生物学学术英语	轻工业,手工业学术英语	人类学学术英语
社会科学学术英语	社会与环境学术英语	神经病学与精神病学学术英语	生理学学术英语	生物工程学(生物技术)学术英语
生物化学学术英语	生物物理学学术英语	石油,天然气工业学术英语	世界文学学术英语	书法,篆刻,雕塑学术英语
数学学术英语	水产,渔业学术英语	水利工程学术英语	特种医学学术英语	体育学术英语

图4-12 "X技术"网的科技英语资源

3.Taking Streaming Current (SO As The Essential Parameter, This Paper Focuses On The Electrokinetic Chemistry Characteristics Of Oxidation And Reduction Process In Water Treatment Condition.以流动电流为参数,研究了水处理条件下氧化还原过程的电动化学特性。

5)Oxidation Reduction氧化还原

1.A New Method,Oxidation Reduction Titration Method,That Is Used In Quickly Analyzing NO 3 - Content Was Put Forward.本文提出了一种快速分析NO3- 的新方法———氧化还原滴定法,适用于硝酸盐生产过程中的控制分

6)Oxidize-Deoxidize氧化-还原

1.One Of The Specimen Went Through Oxidize-Deoxidize Treatment,The Layer Structure Appeared In The Surface,And Then Two Specimens Were Carboned And Quenched Under H_2 Atmosphere In The Resistance Furnace.将其中一试样通过氧化-还原处理,使其表面具有层状结构,与另一脱碳试样一起放入改造后的电阻炉内,在通氢气的气氛中进行渗碳淬火。

2.Sample A Whose Surface Has Layer Structure Via Oxidize-Deoxidize Treatment And Sample B Which Is Decarbonization Are Put Into Resistance Furnace For Carbonization And Quencher Under Hydrogen.将其中试样A通过氧化-还原处理,表面具有层状结构,与另一脱碳试样B一起放入改造后的电阻炉内在通氢气的气氛中进行渗碳淬火。

延伸阅读

氧化还原氧化还原反应即电子的得失。在初中阶段：讨论仅仅是得失氧的问题在高中阶段：讨论即为电子的得失。判断一个化学反应中元素是被还原，或者是被氧化了，即应看其是失去了电子，还是得到了电子。失去了电子即为还原剂，其化合价升高。得到了电子即为氧化剂，其化合价降低。在产物中：被氧化的元素叫做氧化产物被还原的元素叫做还原产物这即使氧化还原。

图4-13 “X技术”网术语双语语料

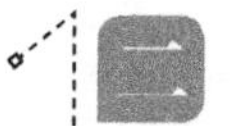

单语语料库

单语语料库为英汉双语译者提供词语搭配、惯用表达等,可以作为词典用法的真实案例补充。译者可使用的英语语料库主要有杨百翰大学语料库(http://view.byu.edu,目前用户量最大的英语语料库)、美国当代英语语料库(Corpus of Contemporary American English,简称COCA,https://www.english-corpora.org/coca/)、美国国家语料库(American National Corpus,简称ANC,https://www.anc.org/)、英语国家语料库(British National Corpus,简称BNC,http://www.natcorp.ox.ac.uk/)等。可使用的汉语语料库主要有语料库(http://yulk.org/)、北京语言大学现代汉语语料库(BLCU Corpus Center,简称BCC,http://bcc.blcu.edu.cn/)、北京大学中国语言学研究中心语料库(Center for Chinese Linguistics PKU,简称CCL,http://ccl.pku.edu.cn:8080/ccl_corpus/)、语料库在线(http://corpus.zhonghuayuwen.org/)等。

第四节　语言问答互助工具

有些术语用法、专业知识等通过各种途径均不容易找到跨语言的对应表述。译者可通过一些大型问答平台搜索答案或临时提问。常用的平台有 Quora（https://www.quora.com/）、HiNative（https://hinative.com/）、ChatGDT（https://chat.openai.com/）等网站以及“我在 AI”手机应用软件。HiNative 是 Lang-8 推出的一款全球语言问答平台，与其他语言交换类社区平台不同的是，该平台更加注重语言问题的问答。HiNative 已经把大多数可能会提出的问题都列出来了，用户只需要填入相应的词或句子即可。

Quora 是一个问答社交网络服务（SNS）网站，用户可通过该网站分享和获取知识。该网站曾采用邀请制，而现在用户通过 Google 或者 Facebook 账号即可登录，既能防止搜索引擎索引内容，又让大众都可以参与。对汉英译者而言，其最大的好处就是可以搜索外语疑难点，找到英语母语高水平人士的回答。同时，还可以向行业人士提问，加深对专业词汇甚至整个文本的理解。搜索结果界面会将问题中的词用底色析出，帮助译者快速筛选有用答案。如翻译建筑行业文本时，若难以辨析“foundation”和“footing”，则可通过输入“difference between foundation and footing”获得如图 4-14 所示结果。

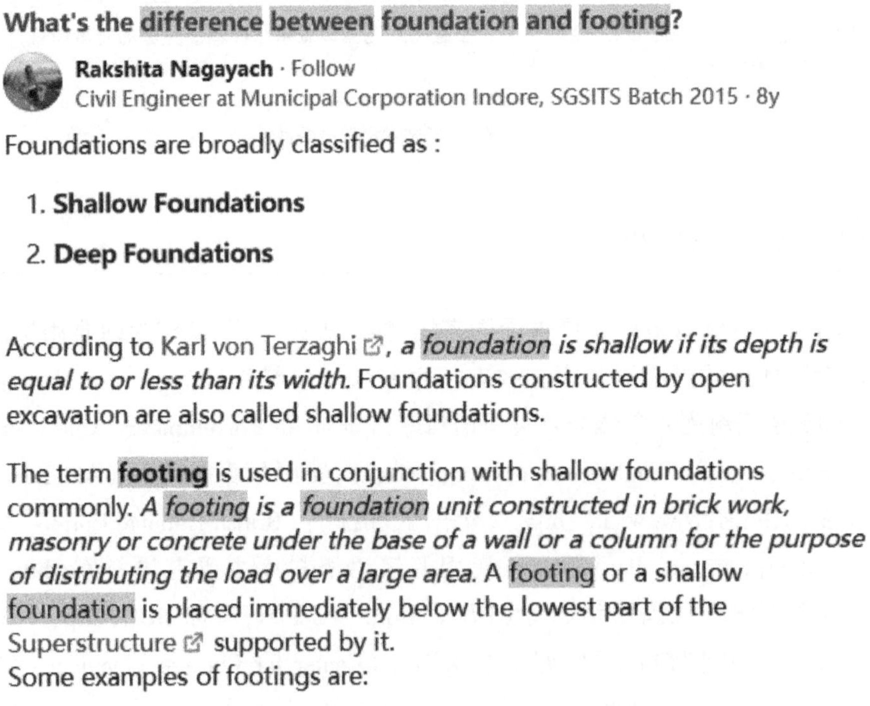

What's the difference between foundation and footing?

Rakshita Nagayach · Follow
Civil Engineer at Municipal Corporation Indore, SGSITS Batch 2015 · 8y

Foundations are broadly classified as :

1. **Shallow Foundations**
2. **Deep Foundations**

According to Karl von Terzaghi, *a foundation is shallow if its depth is equal to or less than its width.* Foundations constructed by open excavation are also called shallow foundations.

The term **footing** is used in conjunction with shallow foundations commonly. *A footing is a foundation unit constructed in brick work, masonry or concrete under the base of a wall or a column for the purpose of distributing the load over a large area.* A footing or a shallow foundation is placed immediately below the lowest part of the Superstructure supported by it.
Some examples of footings are:

图 4-14　“What's the difference between foundation and footing”在 Quora 中的搜索结果截图

该解答由专业领域建造师提供,不仅深入解释了两者之间的区别与联系,还给出了图片对比和具体定义:A foundation is the part of the structure which is in direct contact with the ground to which the loads are transmitted, while the lowermost portion of the foundation which is in direct contact with the sub-soil is called the footing。

ChatGPT(Chat Generative Pre-trained Transformer)是美国 OpenAI 公司研发的聊天机器人程序。ChatGPT 是人工智能技术驱动的自然语言处理工具,能够理解并学习人类语言,与使用者互动对话等。对译者而言,除输入指令让其翻译文本外,最重要的功能是询问专业技术难题和语言使用问题。如翻译“中国天眼”相关文本时,直接在 ChatGPT 页面询问其英语名及相关数据,可获得图 4-15 所示结果。

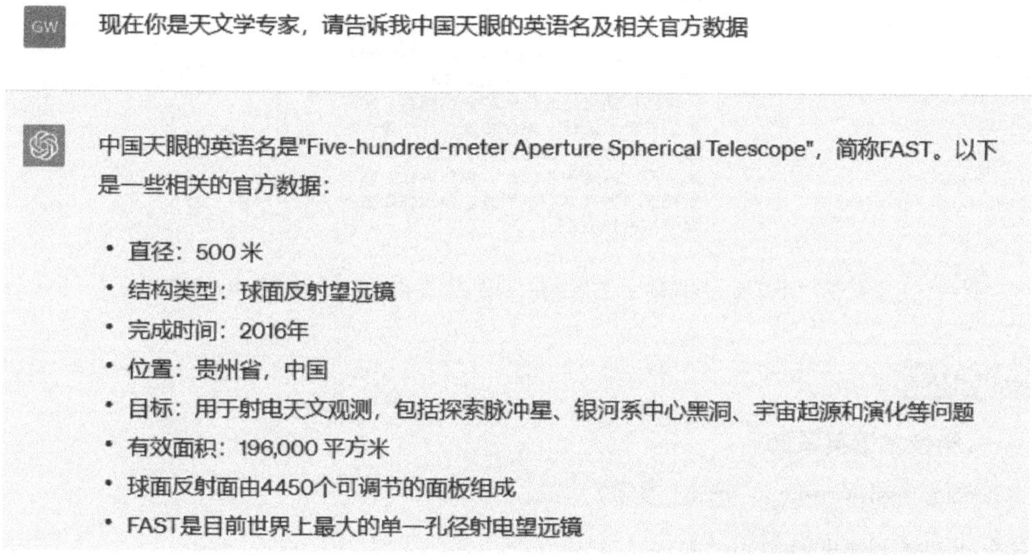

GW 现在你是天文学专家,请告诉我中国天眼的英语名及相关官方数据

中国天眼的英语名是"Five-hundred-meter Aperture Spherical Telescope",简称FAST。以下是一些相关的官方数据:

- 直径:500 米
- 结构类型:球面反射望远镜
- 完成时间:2016年
- 位置:贵州省,中国
- 目标:用于射电天文观测,包括探索脉冲星、银河系中心黑洞、宇宙起源和演化等问题
- 有效面积:196,000 平方米
- 球面反射面由4450个可调节的面板组成
- FAST是目前世界上最大的单一孔径射电望远镜

图 4-15 “中国天眼”英语名及官方数据在 ChatGPT 页面询问结果截图

“我在 AI”是由南京有零科技有限公司研发的一款手机应用软件,契合当前智能 AI 时代的发展步伐,为用户提供全方位、个性化的聊天机器人服务。该软件中各人工智能机器人聊天功能相互独立,因此无须专门输入指令,直接提问即可。回答形式是语音与文字同时出现,相较 ChatGPT 而言,更能满足使用者的听觉需求。例如,在其“我在”或“Chat 5.0”板块中提问“半月板一度损伤”英语表达,索要相关学术文献,咨询相关治疗事宜等,可得到如图 4-16 所示的结果。

本章介绍了科技翻译工作者在遇到术语、表达等难题时能够用到的一些在线工具及相关查询方法,希望能够给译者及翻译学习者以启迪思考。由于篇幅限制且类似网站经常更新,因此未能囊括所有工具,翻译工作者或学习者只有时刻关注新动向,掌握新技术,才能在学习与实践过程中不断提高翻译效率和翻译质量。

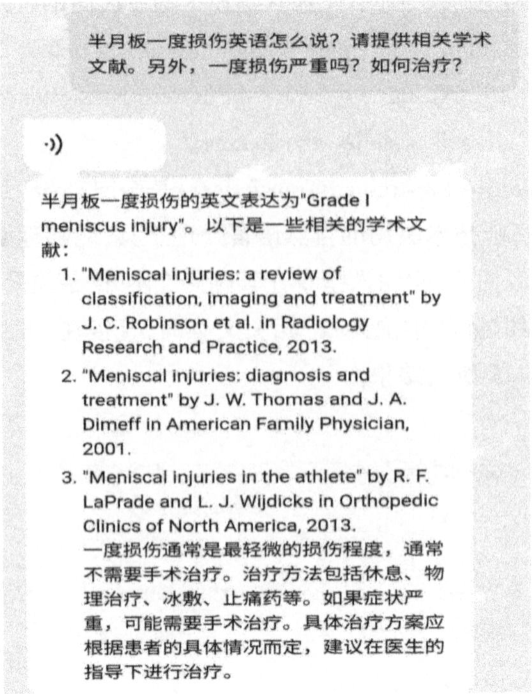

半月板一度损伤英语怎么说？请提供相关学术文献。另外，一度损伤严重吗？如何治疗？

·))

半月板一度损伤的英文表达为"Grade I meniscus injury"。以下是一些相关的学术文献：

1. "Meniscal injuries: a review of classification, imaging and treatment" by J. C. Robinson et al. in Radiology Research and Practice, 2013.
2. "Meniscal injuries: diagnosis and treatment" by J. W. Thomas and J. A. Dimeff in American Family Physician, 2001.
3. "Meniscal injuries in the athlete" by R. F. LaPrade and L. J. Wijdicks in Orthopedic Clinics of North America, 2013.
一度损伤通常是最轻微的损伤程度，通常不需要手术治疗。治疗方法包括休息、物理治疗、冰敷、止痛药等。如果症状严重，可能需要手术治疗。具体治疗方案应根据患者的具体情况而定，建议在医生的指导下进行治疗。

图4-16 “半月板一度损伤”相关信息在“我在AI”手机应用软件中的搜索结果截图

练习题

一、科技术语英译汉

1. compressive residual stress(力学)
2. surface plasmon(材料学)
3. Phaffia rhodozyma (生物学)
4. muscle biopsy(医学)
5. linked queue(计算机科学)
6. envelope instability(物理学)
7. tool JTO. D(石油钻井行业术语)
8. beam pad(电梯行业术语)
9. plain round bars (土木工程行业术语)
10. sedimentary rock(地质学)

二、科技术语汉译英

1. 酶切技术(生物学)
2. 构造煤(地质学)
3. 时空特性(化学)

4. 相选择再结晶(材料学)
5. 镶边空泡(神经病学)
6. 磁抽运(物理学)
7. 绝缘接头(电气工程行业术语)
8. 大洋环流(海洋学)
9. 镗床(制造业术语)
10. 光电功能材料(光电工程行业术语)

第二部分

科技文体的解构与重构

第五章
增译与省译

汉语与英语文化背景差异显著,语言表达侧重点也不同。因此,在翻译科技文本时需要适当分解、调整原文的词语、句型、内容等。译者需要在确保原文意思的基础上,在译文中适当地增加或省略原文的部分短语和句子成分,从而提升译文的流畅性、完整性。基于此,本章将对科技翻译常用的两种方法(增译法和省译法)加以介绍。

第一节 增 译 法

汉语和英语国家的人们有着不同的思维方式、语言习惯和表达方式。因此,在翻译科技文本时,译者需要根据不同的情况合理增添一些原文没有的表达,以便更准确地译出其含义,这样的方法叫作增译法或增词法。增译包括语法增译和内容增译。语法增译指在翻译时增加恰当的表达以保证译文语法正确;内容增译是指译出原文暗含的内容,包括逻辑、文化背景知识以及作者的真实意图等,以使译文更加顺畅易懂。本节将从词汇增译、句法增译和文化增译三个方面举例说明增译法在科技文本翻译中的应用。

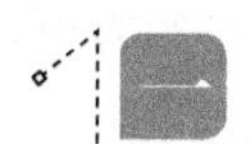

词汇增译

作为句子的组成单位,词决定了句子含义。翻译时可以增加名词、动词、形容词、代词等,以保证译文准确、清晰、完整、流畅。

(一)动词的增译

英语中静态语言居多,常使用名词或名词性短语来表达,而汉语多为动态语言,频繁使用动词,采用大量连动句式。因此,在科技文本英译汉的过程中,可根据语义在名词前增加与该名词搭配的动词,使得译文更加通顺流畅。

(1) Since the early 1970s the dominant theory of what is going on has been that animals learn by trial and error.

译文一 自20世纪70年代初以来,关于这种因果关系存在一种主要理论,动物的学习行为是通过"尝试与错误"实现的。

译文二 自20世纪70年代初以来，关于这种因果关系存在一种主要理论，认为动物的学习行为是通过"尝试与错误"实现的。

分析 如果按照字面意思翻译，译文一无明显问题，但并不符合汉语偏动态的特点。译文二中增加了动词"认为"，更好地表明下文是上文提到的"主要理论"中的观点，更贴近汉语的表达习惯，使得译文读起来更加通顺。

(2) The next stage of space travel is a space station.

译文一 宇宙飞行的下一步是空间站。

译文二 宇宙飞行的下一步是建(立)空间站。

分析 原句是"主语 + 系动词 + 表语"的基本句型结构，表语部分为"a space station(空间站)"。译文一采用直译的方法，将句子直接译为"宇宙飞行的下一步是空间站"，静态表达并不能完整地传达原句的含义。译文二考虑到名词的动态表达，增加动词"建(立)"，再现原句的完整含义，且顺应汉语表达习惯。

(3) The headstock and the tailstock centers must be kept in good condition, or the work will not be accurate.

译文一 必须保养好头架顶尖和尾架顶尖，否则工件达不到应有的精度。

译文二 必须养护好头架顶尖和尾架顶尖，否则加工出来的工件难以达到所需精度。

分析 "the work"指的是"工件"，是在机器(磨床)上加工的产品部件，而头架和尾架都是磨床的重要组成部分。译文一直接将"the work"译为"工件"未能体现其工作性能，与句子的前半部分意义脱离。译文二中添加了动态表达，在"the work"前增译了"加工"这一述语(动词)和"出来的"这一补语，使得句子前后关系明确清晰，更能让读者理解该句中机器的运行原理和作用。

描述物质属性时，汉语多将属性前置、形容词后置，省去谓语动词；而英语必须有谓语动词予以支撑。因此，在汉译英时应适当添加谓语动词，以使译文符合英语表达。

(4) 赤铜硬度大，密度高。

译文 Red copper possesses high hardness and high density.

分析 "硬度大，密度高"是赤铜的主要特点，原文中的汉语表达不需要动词，但翻译为英文时，应在译文中增添动词"possess"，使其符合英语的语法特点。

(5) 目前通过市场监测的齿轮磨床的工作原理基本相同，但型号不同。

译文 At present, gear grinders through market monitoring perform basically the working principle but appear in a variety of models.

分析 按照汉语原文，该句中并不存在动词，在汉译英时可以增加动词"perform"(表现出基本相同的工作原理)和"appear"(在型号方面呈现出不同)，使译文句子完整，更容易让译文读者理解。

(二)名词的增译

英语中多抽象名词,句子通常比较简短,若汉译时采用直译,容易产生词义模糊、歧义等问题。因此,在英译汉时应适当增补名词,将抽象名词具体化,完整流畅地表达句子含义。

(1)Action and reaction are always equal and opposite.

译文一 作用力和反作用力总是相等和相反。

译文二 作用力和反作用力总是大小相等、方向相反。

分析 译文二将"equal"译成"相等"的同时,在其前补充了名词"大小";将"opposite"译成"相反"的同时,补充了名词"方向",使得汉语译文指向性更明确,前后对仗工整,句意准确完整。

(2)The gases produced by the burning fuel rise because they are driven upward by colder and denser air.

译文一 燃烧燃料所产生的多种气体为更低、更大的空气向上驱赶而上升。

译文二 燃烧燃料所产生的多种气体为温度更低、密度更大的空气向上驱赶而上升。

分析 译文一中"更低、更大"指向性不明确,很难让译文读者直接掌握句子含义。译文二在colder"更低"前增补了名词"温度",在denser"更大"前添加了名词"密度",使得翻译内容翔实,指向明确。

为了将原文的含义完整清晰地表达出来,在汉译英时可以根据句意在介词短语中增补表示度量、用途、形式等名词。

(3)化学反应可以用文字表达,但更多的是用化学方程来表达。

译文 Chemical reactions can be stated in words, but more often in the form of chemical equations.

分析 汉语原文采用的表达是"用……来表达",根据原句含义,原文强调的是化学反应的"表达形式"。因此,在英译时增加名词"the form",强调"化学方程式"这一表达形式,使得句意完整流畅。

(4)金属经过热处理后,强度更大,更加耐用。

译文一 Going through the heat treatment, metals become much stronger and more durable.

译文二 Going through the process of heat treatment, metals become much stronger and more durable.

分析 译文一将"热处理"处理为"the heat treatment",在词义上对等,但却忽略了"热处理"是一个过程性操作的科技事实。译文二则为"热处理"增译了名词"……过程"译为(process of...),将"热处理"的过程性操作体现完整,更有助于读者对科技文本的理解。

(三)形容词和代词的增译

英语中常用名词来表达物质的功能、性能、使用特点等,用词准确精简。因此,汉译时,应结合上下文,在名词前增添合适的形容词,让译文更加符合原文含义。

(1) Performance and accuracy are the main advantages of automatic machine tools.

译文一 性能和精度是自动机床的主要优点。

译文二 性能强、精度高是自动机床的主要优点。

分析 译文一直译了原文中的两个名词"performance"和"accuracy"为"性能"和"精度",缺少形容自动机床优点的词语。因此,译文二在"性能"后增加形容词"强","精度"后增加形容词"高",更为直观地表明自动机床的特点。

英译时,要适当增加形容词性物主代词,表明所属关系,符合英语表达习惯,使得句意更加完整。

(2)一个原子本身就是一个完整的整体,包括电子、质子、中子及其他元素。

译文一 An atom itself is a complete whole, with electrons, protons and neutrons and other elements.

译文二 An atom itself is a complete whole, with its electrons, protons and neutrons and other elements.

分析 根据原文含义可以分析出"电子、质子、中子及其他元素"是"原子"的一部分,属于"原子"本身。译文一采用直译方式,指向不够明确,导致句子逻辑关系不清。译文二增加形容词性物主代词"its",表明所属关系,明确句子含义。

(3)电鳐能放出强大的电流把猎物击晕。

译文 Electric rays stun their prey with huge electrical discharges.

分析 根据原文含义"电流"在本句中指的是由"电鳐"本身释放出来的物质,属于"电鳐"本身而非其他生物。因此,在翻译时增加形容词性物主代词"their"表明所属关系。

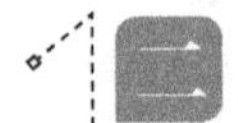

句法增译

科技翻译除了需要注意词的翻译外,还应注意句法结构的准确性。在翻译时可以根据上下文逻辑关系、汉语和英语的表达习惯,增译主语、谓语或其他成分。

(一)主语的增译

根据句法结构,英文中常使用抽象名词、指示代词等作为主语,英译汉时则需要增译主语将抽象名词具体化,使得译文表达更清晰,避免歧义、错误。

(1) This accumulation is due to human causes. Excess gas will cause excessive atmospheric pressure.

译文一 这种积累归咎于人为因素。多余的气体会使大气压力增大。

译文二 这种气体的积累归咎于人为因素。多余的气体会使大气压力增大。

分析 原文中主语“accumulation”意思是“积累”，译文一采用直译的方法，忽略了上下文的句意关系，致使其意思略显模糊。译文二在主语“accumulation”前根据下文增补形容词“气体的”译为“气体的积累”，既排除了歧义，也较好地传达了原义。

(2) There are substances through which electric currents will not pass at all.

译文一 有物质电流根本不能通过。

译文二 许多物质根本不让电流通过。

分析 原文中主语“substances”（物质）为英语的名词复数形式。译文一直接将其译为“物质”，只译出主语本义“物质”而忽略了主语的复数形式，表达不准确。译文二将原文主语的意义与形式相结合，在主语“物质”前增加形容词“许多”，准确表达出原文的含义。

(3) Living things are composed of cells.

译文一 生物都由细胞组成。

译文二 所有生物都由细胞组成。

分析 原文“living things”为复数形式。译文一将其译为“生物”没有将主语的复数形式完整译出，容易产生歧义。结合生物常识和主语的形式，译文二将主语“生物”增译为“所有生物”，明确表达范围和含义。

科技翻译注重清楚、准确、精炼、简洁。面对汉语省略主语的情况，在汉译英时要根据上下文增补对应的主语成分，让英文句子准确无误。

(4) 为释放出电子，必须使其高速旋转以摆脱原子核束缚。

译文 To free electrons, something has to make them whirl fast enough to break away from their nuclei.

分析 原文为汉语无主语句。在英译时需要结合科技文本上下文，增加适当的主语成分保证句法结构的完整。本句为肯定句，在主语不明确时，可以增加不指明代替任何特定名词或形容词的不定代词“something”做主语，让句子结构完整。

(5) 有了金属探测器之后，几个人花几个月的时间就可以完成了。

译文 With the metal detector, the work can be completed by several people in a few months.

分析 根据原文句意和句法结构可发现，“几个人花几个月的时间就可以完成”的是“某项特定的工作”。因此，在翻译时增加原句中省略的主语“工作”（译为“the work”），明确句子主语，使原文内含传递更加完整。

（二）谓语的增译

英语中出现两个或多个相同谓语时，常常会省略，以使句子表述简洁不赘余，而中文中则需将其译出，使得表达更加地道。英文中的介词短语译为中文时往往要增译谓语，使句法结构完整，表达更加明确。

(1) Oil is obtained more easily than coal.

译文一 油比煤被获得得容易。

译文二 采油比采煤容易。

分析 原句为英语的被动句式。译文一保留了原文的句式,直译为"油比煤被获得得容易"。很明显,这样的直译使得句意不明确,且句子拗口不通顺。因此,翻译此类句型时,要适当增译谓语动词。译文二增译谓语"采",将"油"和"煤"两个名词增译谓语为两个动作"采油""采煤",明确动作实施对象,表达完整,符合原文含义。

(2) Due to the lack of time left, we completed the first group of experiments in the first week, and the second group of experiments the next week.

译文 由于所剩时间不足,我们第一周完成了第一组实验,第二周完成了第二组实验。

分析 英语句子表达通常比较简洁。同一个句子前后谓语相同时,有时会省略后一个谓语。在翻译时,译者需要将省略的谓语补充完整,明确句意。原句中省略了谓语"completed",因此在翻译时需增添谓语"完成",符合汉语表达方式,明确句意。

一个完整的汉语句子可以没有谓语成分,然而在英语句法结构中,谓语却是不可或缺的。因此,在汉译英时应根据句子类型增添恰当的谓语,使英语句子正确、完整。

(3)标准化零件可以通用。

译文 Standardized fittings can be employed universally.

分析 在汉译英时要注意主动句和被动句之间的转换。本句中汉语原句为主动句,在翻译时需要根据英语表达习惯,将其译为被动句。因此,译文中应增添谓语"employ"的被动态。

(4)这种激光束的频率范围很窄。

译文 This laser beam covers a very narrow range of frequencies.

分析 原文在描述"激光束"的频率范围特点。汉语在描述事物特点时,一般使用相应的形容词来修饰主语。但在英文中,为了保证句法的准确以及句意的明确,会使用谓语动词表达事物某方面的特质。该译文中就增译了谓语动词"cover",在保证句子成分完整的基础上,生动形象地描述出"激光束"的频率范围特点。

(三)省略成分的增译

为使表达简练,英语中经常会出现句子成分省略的情况。而翻译时,则需要根据理解增补原文缺失的词义。英语最忌讳重复,前面出现的部分一般后面不再出现,以避免重复。但汉译时往往需将其增补上,使得汉语句子完整,句意明确。

(1) Like charges repel each other, but opposite charges attract.

译文 同性电荷相互排斥,异性电荷相互吸引。

分析 原文中后半句省略了“each other”,在汉译英时需要补充完整,符合汉语表达,使句子对仗工整,含义明确。

(2) As we know, all matter consists of molecules, and molecules of atoms.

译文 众所周知,一切物质由分子组成,而分子则由原子组成。

分析 根据原文句意,句子后半句省略动词“consist”,在翻译时需要补充完成,明确句子的所属关系和组成关系。

(3) A generator may generate electricity when driven by a steam engine, gas engine, or water wheel.

译文 当发电机由蒸汽机、燃气轮机或水轮机驱动时,它便能发电。

分析 原句为条件状语从句,英文表达省略了从句的主语,翻译时需要补充缺失内容,使句子逻辑清晰,表达明确。

三 文化增译

科技翻译主要涉及呈现客观事实的内容。基于对读者友好的理念,文章中的术语、专有词语、公式、算法等第一次出现时,有时需要增译出其背景信息。同时,在英译汉时可以恰当地增译成四字词组,使译文的表达更精练,句式更协调,语义更深远,既起到修辞作用,又丰富文化内涵,同时大大增加科技类文章的可读性。

(一) 背景信息的增译

科技翻译虽注重简洁,但随着科学技术的发展和普及,应该让更多非科技领域的读者理解科技类文章,并通过阅读此类文本获得科技知识。因此,翻译时需要对专有名词、实验过程、科技术语等重要背景信息做补充或解释说明,以此打破学科壁垒,让读者更容易理解句子含义。

(1) Mosquitoes follow carbon dioxide (CO_2) gas because it is breathed out by animals, on whom they feed transmitting diseases such as malaria and dengue in the process.

译文 蚊子追踪二氧化碳(CO_2)气体,因为它们叮咬的动物(此处指人类)呼出这种气体,蚊子在叮咬过程中传播诸如疟疾和登革热等疾病。

分析 原文中省略了“蚊虫叮咬的动物”的所属范围,在翻译时需要结合上下文含义,给原文中缺失的、容易造成阅读障碍和困难的部分补充背景信息。本句译文结合句意,增补了蚊虫叮咬的动物“此处指人类”,其所指范围得以明确,从而避免了歧义的产生。

(2) Based on how the physical forms embody the three contents, and by the methods of AHP and Delphi, this paper sets up the indices system for evaluating the ecological features of environment in residential areas.

译文 围绕着住宅区环境的各种物质形式如何体现三方面内容，采用 AHP 法（层次分析法）和 Delphi 法（德尔斐专家调查法），构建起住宅区环境生态性评价的指标体系。

分析 AHP（全称为 Analytic Hierarchy Process），意为"层次分析法"，Delphi 音译为"德尔菲法"，又称"（德尔菲）专家调查法"，旨在起到一定的释义作用。通过此类增译，译文读者可知二者为实验方法，进而明白原文大意。

（3）The cost of scaling this up is unknown, but will probably be far more than the 145 pound that a normal blood donation currently costs in Britain.

译文 扩大生产规模的成本尚不明确，但可能会远远高于英国目前正常输血的 145 英镑（约人民币 1125 元）的成本。

分析 由于文化的差异，在翻译科技文本中的度量单位（如长度单位、重量单位、货币单位）时，需要根据目的语的表达习惯，在括号中增译出其信息。本句中将"145 pound"译为汉语时，增译了"约人民币 1125 元"，贴近译文读者的文化背景和数字概念，将信息快速准确传达。

（二）四字词组的增译

从文化层面和读者友好角度，将科技英语文本翻译为汉语时，可以增译成四字词组，使得句子在准确清晰的基础上更具文学性和可读性。

（1）Cutting grinders vary in design from simple machines having a limited purpose to complex universal machines that can be adapted to any cutter grinding requirement.

译文 工具磨床结构多种多样，从简单且用途有限的工具磨床到复杂的、能满足任何工具磨削要求的万能工具磨床，种类繁多，不胜枚举。

分析 原句中"vary in design"表明了工具磨床结构的多样性，在下文又使用"machines"名词复数表达，强调了机器数量多、种类多。因此，在翻译时增译两个四字词组"种类繁多"和"不胜枚举"，既表明了句意又增强了语气。

（2）Larger tests, including tests on actual patients, will be needed before this practice.

译文 将这一方法付诸实践之前，需要开展更大规模的测试，包括在真实患者身上做测试。

分析 原文中"practice"意思是"实践"。根据原文句意，"实践"指的是该方法最终投入使用。因此，为了将句子含义清晰明确地表达，可将"实践"增译为四字词组"付诸实践"，以此清晰简捷地表达出原文句意。

（3）To ensure the accuracy of experimental data, every link in the experiment is crucial.

译文一 为保证实验数据准确，每个实验环节都是重要的。

译文二 为保证实验数据精准无误，每个实验环节都至关重要。

分析 译文一对原文中"accuracy"和"crucial"两个词采用了直译的方法，译文二采用了增译法，将两个词语增译为两个四字词组"精准无误""至关重要"。两者相比，译文二更加通顺,突出强调实验数据和环节的重要性。

第二节 省 译 法

省译法又称减译法或减词法,与增译法相对,指省去不符合目的语思维习惯、语言习惯和表达方式的词,以避免译文累赘。在中英互译的翻译方法研究中,省译大多都是以避免译文冗余、累赘为目的,调整句法和语法,省略某些词以求译文更加简洁、流畅、通顺。与增译法一样,在翻译时,何处应“增”,何处应“省”,这一点十分考验译者的能力。在翻译科技文本时,译者不仅需要有良好的中文功底,而且还要对相关领域的科技术语、目的语的用法习惯有较深的了解,才能做到忠于原文基础上的精准省译。本节将从词汇省译和内容省译两方面,举例说明科技翻译中如何使用省译法。

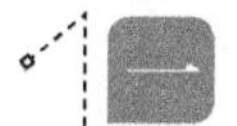

一 词汇省译

科技文本具有严谨简洁的特点,一般翻译此类文本时不允许删减原文内容。但是,汉语和英语中词汇的使用方法差异很大,为了保证科技文本翻译的准确性,在翻译时可以适当省略名词、动词、代词、连词等词语,可使得译文完整流畅。

(一)代词的省译

英语中经常使用各类代词来指代上文中提及的事物,种类丰富,用法复杂多样。与英语相比,汉语中的代词种类较少,用法较为简单。在翻译科技文本时,适当省略代词可以让译文更加简洁、准确。

(1) Under appropriate conditions, put one or more oxygen atoms together with another chemical element and you will have oxides.

译文一 在恰当的条件下,将一个或多个氧原子与另一种化学元素结合你能够得到氧化物。

译文二 在恰当的条件下,一个或多个氧原子与另一种化学元素结合能形成氧化物。

分析 如果根据原文逐字翻译,无疑就会得到译文一。但十分明显的是,原文第二人称代词“you”翻译成“你”稍显冗余。因此,翻译科技文本时,我们应该追求简洁、准确,即用精确的语言表达原文的含义。译文二中对原文的“you”做了省译处理,使得译文读来更加通顺。

(2) DNA carries the genetic blueprint which tells any organism how to build itself.

译文一 脱氧核糖核酸带有决定有机体自我的形成方式的遗传型板。

译文二 脱氧核糖核酸带有决定有机体形成方式的遗传型板。

分析 原句中的反身代词"itself"指代句子中的"DNA"。译文一将"itself"译为"自我的",在译文中作形容词修饰"形成方式",此类翻译将简单句复杂化,指代不明会生出歧义。译文二将"itself"省译,在保持原句含义的基础上,让译文前后逻辑更清晰,句意表达简洁明了,通顺完整。

(3) The facility itself consists of two capture plants fitted with more than 4,000 instruments to monitor what is going on, and with a total capacity of 80,000 tonnes of carbon a year.

译文一 这个装置自身由两个收集设备组成,这两个设备里面安置了4000个用来监控设备运行的仪表,每年能收集8万吨碳。

译文二 这个装置由两个收集设备组成,这两个设备里面安置了4000个用来监控设备运行的仪表,每年能收集8万吨碳。

分析 原句中的反身代词"itself"指代"the facility"。在英文中使用反身代词是用来强调主语本身,"自身"二字出现在"装置"后略显冗余,背离汉语的书面表达习惯。译文二所做的省译处理,既满足科技文本简洁明确的特点,也符合汉语常见的表达方式。

(4) They analyzed these 40 chemicals that past research has suggested have socioeconomic significance.

译文一 他们分析了这些40种化学物质,过去的研究表明这些物质具有社会经济学意义。

译文二 他们分析了这40种化学物质,过去的研究表明这些物质具有社会经济学意义。

分析 原句中代词"these"指的就是"40 chemicals"。译文一将"these"译为"这些"并置于"40种化学物质"之前,使得译文复杂冗余,不符合科技文本的特点。因此,译文二使用省译法,将"these"处理为"这",在不影响原句含义的基础上,使得译文简洁且顺口。

(5) A carbon sink is anything that absorbs more carbon than it releases.

译文一 碳汇指的是任何事物,它们吸收的碳比释放的碳多。

译文二 碳汇指的是碳吸收量多于碳释放量。

分析 原句中"anything"为不定代词,同时也是定语从句的先行词。译文一采用直译法,将"anything"直译为"任何事物",同时将定语从句关系代词"that"增译为"它们"。总体上看,译文一表达冗余复杂,句子结构不清晰,指代不明确。译文二采用省译法,将不定代词"anything"做省略处理,直接表明主语"a carbon sink"所指的内容,符合汉语表达习惯和科技文本简洁清晰的特点。

(二) 名词的省译

在英文的科技文本中,有时会使用两个意义相同或相近的词语表达来阐述同一个概念,或者为了句子结构的完整,使用一些名词作为句子中不可或缺的成分,但汉语中不常用此类表达方法,这时就需要使用省译法,即省略其中一个,使得译文更加通顺流畅,简洁清晰。

(1) This refers to the fact that it is not necessary to detect their concentration and humidity in this experiment.

译文一 这指的这一事实是不必检测它们的浓度和湿度。

译文二 这指的就是不必检测它们的浓度和湿度。

分析 原句是由关系代词"that"引导的定语从句,从句的先行词为"the fact"。译文一将"the fact"直译为"这一事实",指的就是从句中包含的内容,即"不必检测它们的浓度和湿度",英语中使用"the fact"是为了让句子结构完整,充当定语从句或同位语从句的必要成分,但译为汉语就有些多余。因此,译文二省略"the fact"部分,直接将原句想要阐述的事件本身译出,既简洁也达意。

(2) A complex password is a password that has at least eight characters and consists of uppercase and lowercase letters, as well as numbers or other symbols.

译文一 一个复杂的密码应是一个密码:至少有8个字符,包括大小写字母、数字和其他符号。

译文二 一个复杂的密码应至少有8个字符,包括大小写字母、数字和其他符号。

分析 原句中出现了两个"password",第一个"a complex password"是原句中的主语,必须要翻译完整;第二个"a password"是对主语的再次强调或补充,同时充当定语从句的先行词。译文一将其直译为"一个复杂的密码应是一个密码",不仅略显冗余,而且"搞笑",虽不违背汉语行文习惯,但就科技文本翻译而言,似乎不够严肃或严谨。因此,译文二将原句中第二个"a password"省译,直接将主句和从句部分的语义连接,句意准确,逻辑清晰。

同时,在翻译汉语科技文本中部分主谓短语或相关表达时,通常只需要译出形容词,省略名词,让译文更加符合英文表达习惯,简洁直观地译出原句的含义。

(3)根据科学家的看法,引入外来物种也是为了降低当地物种灭绝的风险。

译文一 According to scientists' opinion, the introduction of exotic species is also to reduce the risk of local species extinction.

译文二 According to scientists, the introduction of exotic species is also to reduce the risk of local species extinction.

分析 原句中"根据科学家的看法",译文一直接处理为"According to scientists' opinion",看似对应,实则有悖英语表达习惯,在不影响原句句意传达的基础上,译文二删除名词"opinion",此举不仅更简洁,而且更地道。

(4)对沉积物的检查表明,即便是最古老的物质中也含有一些细菌。

译文一 Examination of the sediments showed that even the oldest things still contained a few bacteria.

译文二 Examination of the sediments showed that even the oldest still contained a few bacteria.

分析 根据原文句意,原文中"最古老的物质"强调的部分为"最古老的"。译文一采用直译法,将原句该部分译为"the oldest things",信息传递准确。译文二使用省译法,将名词"物质"虚化省略处理,译为"the oldest",既不影响原句句意的表达,在让译文更加简洁的同时,也让重点部分得以凸显。

汉语中有一些用来表示现象、行为、属性等所属范围的名词,本身没有实质意义,翻译成英文时,可以省去不译,让译文更加通顺流畅。

(5)根据水的蒸发现象,人们知道液体在一定条件下能变成气体。

译文一 From the phenomenon of evaporation of water people know that water can turn into gases under certain conditions.

译文二 From the evaporation of water people know that liquid can turn into gases under certain conditions.

分析 原句中的"蒸发现象",译文一将其直译为"the phenomenon of evaporation",译文二省略掉"现象"这一范畴词,译为"the evaporation"。原句是由于汉语多用范畴词概括某些情况,让句子成分更加清晰完整,但英语中的很多名词本身就包含现象、作用、原理、工作等含义,而译文二之所以显得更准确且简洁,正是因为译者较好地把握住了这一要点。

(三)动词的省译

英语和汉语在动词的使用方式上有着较大的不同。汉语句子中动词、形容词、名词、介词等都可以用作谓语动词,然而,英语句子不可以缺少谓语动词。英语的谓语动词种类繁多,其中包含一些词语本义被虚化的动词,如"have, make, get, give, take"等,它们可以和许多词搭配,但本身几乎没有意义,它们的意义主要通过后面的名词来体现。因此,在翻译此类词组时可以省去虚化动词,直接翻译其所搭配的其他词。

(1) Old engines give poor acceleration so to achieve acceptable performance hybrid technology is required.

译文一 老旧的发动机提供较慢的加速,所以需要使用混合动力来达到理想性能。

译文二 老旧的发动机加速较慢,故需用混合动力来达到理想性能。

分析 原句中的动词为"give",与其搭配的名词为"poor acceleration"。译文一将"give"翻译为"提供",没有联系原文句意分析出原句中"give"的虚化情况。因此,译文二将动词"give"省译,更加清晰直观翻译出原句句意。

(2) This disinfectant has the ability to kill a certain number of bacteria, and can play a certain cleaning role in combination with detergent after development.

译文一 这种消毒剂有能力杀死一定数量的细菌,研发之后可以结合清洁剂起到一定的清洁作用。

译文二 这种消毒剂能杀死一定数量的细菌,研发之后可以结合清洁剂起到一定的清洁作用。

分析 原句中的谓语动词及其搭配为“have the ability to…”,其中“have”是被虚化的动词,下文中已经具体陈述了有能力做的具体事件。译文一采用直译就会让译文变得复杂冗余,指向不明。对于此类科技文本,最好的翻译方式就是像译文二那样,省去被虚化的动词,直接阐述主语具备的能力,使译文更加直观清晰。

(3) All one has to do is take a quick look around to appreciate modern society's dependence on the artificial intelligence.

译文一 每个人需要去看一下周边的万事万物,就会发现当今社会已离不开人工智能了。

译文二 看一下周边的万事万物,就会发现当今社会已离不开人工智能了。

分析 原句中的谓语动词及其搭配为“take a quick look around to…”,其中“take”是被虚化的动词,句子的翻译重点应在原句的后半句“modern society's dependence on the artificial intelligence”。译文一将原句直译,比较冗余复杂。译文二将动词“take”省译,将译文的重点放在后半句,详略得当,译文句意准确。

在汉语科技文本中,经常出现动宾搭配的词组,如采取方法/措施、调整位置、定向处理、计算公式等,翻译时通常可以省略其中的动词,将其译成相应的动名词形式或含有动作目的意义的名词。

(4)为获取最高精度,通常用化学方法来进行热测量。

译文一 For getting maximum precision, thermal measurements are usually made by chemical methods.

译文二 For maximum precision, thermal measurements are usually made by chemical methods.

分析 原句中“为获取最高精确”,重点在“最高精确”。因此在翻译时可以将汉语动词“获得”虚化省译。译文一采用直译,译为“For getting maximum precision”,比较复杂冗余,不能直接表达原句强调的重点。译文二则采取省译法,省掉动词“获取”,译为“For maximum precision”,使得句子目的明确,逻辑清晰。

(5)采取自动切割的方法能使切割系统的操作更迅速、更准确,而且更灵活。

译文一 Taking automatic cutting methods allow for greater speed, accuracy and flexibility in the operation of cutting systems.

译文二 Automatic cutting methods allow for greater speed, accuracy and flexibility in the operation of cutting systems.

分析 原文中“采取自动切割的方法”,着重强调的部分是使用的方法而非动词“采取”。译文一使用直译法,将原句中的谓语动词“采取”译为“Taking”,较为多余。译文二使用省译法,在不影响原句句意的前提下,省略动词“采取”,将“Automatic cutting methods”置于句首,使得译文逻辑清晰,重点明确。

(四)连词的省译

英语科技文本的句式和用词严谨准确,更加注重形合,常用连词来建立句子内部的逻辑关系。在翻译成汉语时,要考虑汉语重意合的特点,可以通过省略连词使句子中重点强调部分意义连贯,分析汉语句子中内在的、暗含的逻辑关系。

(1) Open the instruction manual of the instrument, and you will see the product parameters.

译文一 打开仪器的使用说明书,然后你就可以看到产品参数。

译文二 打开仪器的使用说明书,就可以看到产品参数。

分析 原句是由连词“and”连接句子的两个部分,在原文中起到承上启下的作用。但是翻译为中文时要考虑中文的表达方式以及句子内部蕴含的逻辑关系。译文一采用直译的方法,将“and”译为“然后”,在句意层面略显多余,因为“打开仪器的使用说明书”就会看到“产品参数”。译文二将“and”省译,直接翻译为“可以看到产品参数”,明确表达原句含义,同时符合科技文本简洁准确的特点。

(2) Since the process is a mechanical one and does not require heat, it can be very precisely controlled.

译文一 因为整个过程是一个机械加工过程,且无须加热,所以可以精确操控。

译文二 整个过程是一个机械加工过程,且无须加热,因此可以精确操控。

分析 原句是由从属连词“since”引导的原因状语从句,译文一采用顺译法将其译为“因为……所以……”,准确地表达出了句子的逻辑关系,但是表示原因的连接词略显冗余。译文二则使用省译法,省略对“since”的翻译,直接翻译表示原因的前半句,后用“因此”连接原因与结果,译文在保证逻辑清晰的前提下,符合科技文本简洁明确的表达特点。

为了准确传达科技信息和科技要素,汉语科技文本在句与句之间或者句子的各成分之间,有时需要使用一些连接词才能将它们的层次、条理、关系等表达清楚。若在翻译时不需要形式对等,可以使用省译法,省略部分连接词,用英语中特定的语法习惯或用词习惯来表达原文的语言逻辑。

(3)虽然生育能力与年龄有关,但其对不同人的影响不同。

译文一 Though fecundity is tied to age, but the effect of time’s passage varies.

译文二 Though fecundity is tied to age, the effect of time’s passage varies.

分析 原句中采用汉语中常用的表示让步关系的搭配“虽然……但是……”,但在英文中表示让步关系的从属连词不能与“but”连用,因此译文一在语法层面是错误的。对于汉语科技文本中出现的此类句型，翻译时可像译文二那样借助省译法，将“but”省略，保证译文语法准确，句意明确。

(4)由于没有车轴零件，所需维护工作量可减少到最低限度。

译文一 Since there are no axle parts, the maintenance workload can be reduced to the minimum.

译文二 Without axle parts, maintenance requirements are cut to the minimum.

分析 原文是由从属连词“由于”引导的原因状语从句。译文一采用直译法，将“由于”译为“since”，后用“there be”句型译出从句部分，译文比较冗长，没有主要强调内容。因此，译文二采用省译法，省略对从属连词“由于”的对照翻译，使用“without”的复合结构，清晰直观地表明从句部分要强调的成分为“没有车轴零件”。

(5)如果把磁铁靠近某物体，后者就会被吸过去。

译文一 If you bring a magnet near an object, and the latter will be attracted towards it.

译文二 Bring a magnet near an object, and the latter will be attracted towards it.

分析 原句是由“如果”引导的条件状语从句，汉语中一般习惯将假设放在句子前半部分，导致的结构紧随其后。译文一采用直译法，将原文直译为由“if”引导的条件状语从句，译文在语义表达上基本完整。译文二则采用省译法，将从属连词“如果”省略，直接翻译从句中的主要部分“把磁铁靠近该物体”为“Bring a magnet near an object”，相当于使用祈使句强调动作发生的重要性，更加符合英语表达方式和科技文本简洁明确的语言特点。

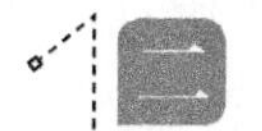

内容省译

科技翻译除了需要注意词语的翻译外，还应注意句意表达和译文内容的准确性。在翻译时可以根据上下文逻辑关系、汉语和英语的表达习惯，省译原句中的重复部分和事理逻辑冲突成分。

(一)重复成分的省译

汉语和英语的词语在词义范围和表达能力方面均存在不同程度的差异。英文复合长句中通常会使用代词、名词、动词等作为句子的重复成分，以此从形式上保证句子的完整，内容上强调句子要突出的主要内容。因此，根据英汉表达方式的不同，在翻译此类科技文本时，应将英语中必要但是汉语中重复或多余的成分做省译处理。

(1) Another benefit of protein is believed to be its ability to maintain normal plasma osmotic pressure and balance of material exchange between plasma and tissue.

译文一 据说，蛋白质的另一个益处是有它的能力去维持正常的血浆渗透压，保持血浆和组织之间的物质交换平衡。

译文二 据说，蛋白质的另一个益处是能维持正常的血浆渗透压，保持血浆和组织之间的物质交换平衡。

分析 原句中的主语为"another benefit of protein"，下文中的物主代词"its"指的就是"蛋白质"，与主语部分重复，但在英文句子表达中合理恰当。译文一使用直译法将物主代词"its"译为"它的"与下文顺译衔接。译文二则采用省译法，将物主代词"its"省略处理，直接译出"蛋白质的益处"，在译文准确的基础上，更加简洁清晰、通顺流畅。

(2) A combination or synthesis reaction results when two or more substances unite to form a compound.

译文一 两种或多种物质合成一种化合物时，就产生结合或者合成反应。

译文二 两种或多种物质合成一种化合物时，就产生合成反应。

分析 原句是由"when"引导的时间状语从句。主句的主语是"a combination or synthesis reaction"，由"or"连接的两个词语表达的是同一含义，即"合成反应"，原句使用这种表达方式可以让目标读者更理解合成反应的含义。但如果像译文一那样直译，就会出现冗余的情况，译文不通顺，表达不准确。因此，在遇到此类情况时，应像译文二那样采用省译法，将重复的名词部分省译，译为"合成反应"，简洁明确，句意清晰。

(3) Smart or intelligent materials have the ability to adapt to their surroundings.

译文 智能材料能适应周围环境的变化。

分析 原文中"smart or intelligent"都表示"智能的"，翻译为汉语时需要采用省译法，省去其中一个形容词的翻译，才能使得译文准确，句意清晰。

与英语相比，汉语更加注重句式的平衡、押韵，以及句子之间的逻辑关系，常使用排比、对仗、重复等修辞手段。在翻译此类科技汉语文本时，要恰当地使用省译法，让译文语言简练、逻辑清晰。

(4) 现已证明，这种材料比任何一种金属都更适合于这类用途，而且其价格也比任何一种金属的都低得多。

译文 It has been proved that this material is more suitable for the purpose than any metal, and also that the cost is much lower.

分析 原文为使句子前后对仗工整，句中重复出现了两个"任何一种金属"，在翻译时可以使用省译法，将第二个"任何一种金属"用代词"that"代替，不会造成歧义，使得译文简洁明确、逻辑清晰。

(5) 若要判断一个物体是否会在水中浮起，就得知道该物体的密度比水大还是比水小。要是密度比水大，该物体会下沉，要是密度比水小，就会浮起。

译文 When trying to decide whether an object will float in water, you need to know whether its density is greater or less than that of water. If it is greater, the object will sink; if less, the object will float.

分析 原文由两个句子组成,其中包含三个"密度",指的都是"物体的密度"。第一个句子作为第二个句子发生的条件,在翻译时可以省略第二处和第三处的"密度",用代词"it"代替,或直接在条件状语从句中使用"if + adj.(less)"结构。在不影响译文句意的基础上,更加符合英文表达,简洁明确。

(二)事理逻辑的省译

在描述实验过程、阐明科学原理、探讨科学问题时,汉语科技类文章里常会使用一些特定的表达,如"由于……的缘故""虽然……但是……""提高……的准确率"等,这是陈述事理逻辑时必不可少的汉语固定搭配。但是,英语中没有与其直接对应的结构或相同含义的表达,翻译时需要采用省译法,使得译文语言表达在准确传达原文句意的基础上,符合英文表达习惯,做到逻辑清晰、通顺流畅。

(1)由于经费有限,我们不能对这个实验做进一步分析。

译文一 Due to limited funds, we cannot make further analysis of this experiment.

译文二 Money does not allow us to analyze this experiment further here.

分析 原文是由"由于"引导的原因状语从句,阐明不能对实验做进一步分析的原因。译文一采用直译法,按照句子顺序逐字翻译,但译文较长,不够简洁。译文二则采用省译法,将"由于"省译,直接将问题出现的主要原因"Money"呈现在句首,如此一来,译文的表述相对更简洁明了。

(2)利用视频电话,不仅可以听到通话人的声音,还能看到人。

译文一 Using a visual phone you can not only hear the person you are talking to, but also can see him.

译文二 With a visual phone you can not only hear but also see the person you are talking to.

分析 原文中为了说明视频电话的功能,采用"不仅可以……还能……",英语中的对应表达为"not only…but also…",译文一采用直译法,按照句子顺序将信息全部译出,译文中出现了两个情态动词"can",比较冗余。译文二采用省译法,将原文中的能愿动词"可以""还能"省译,并用"with"的复合结构开头,译文详略得当,句意明确。

(3)因为使用3D自动打印机,工厂的产品比以前便宜多了。

译文一 Because of the use of 3D automatic printers, the factory's products are much cheaper than before.

译文二 A 3D automatic printer has made the products of manufacturers very much cheaper than before.

分析 原句中用“因为……”这一结构来表明句子的因果关系。译文一采用直译法，将“因为……”译为“because of”，与译文二采用省译法将主语“3D 自动打印机”置于句首相比，译文一显得成分冗余复杂，译文二的表达更加符合英文科技文本表达特点，译文逻辑清晰、简洁通顺。

练习题

一、英译汉

1. The next stage of the project, to be carried out this month, is to take 50 grains from each of the successful plants and repeat the process with them.

2. An element's atomic number is the number of protons in its nucleus.

3. His evidence comes from electromagnetic surveys carried out on some 40 volcanoes, including Mount Fuji in Japan, Mount St Helens in America and others in Bolivia, New Zealand, the Philippines and elsewhere.

4. There might be as much as 1.4m tonnes of copper beneath New Zealand's White Island volcano, whereas the world's largest mines hold tens of millions of tonnes of it.

5. A combination or synthesis reaction results when two or more substances unite to form a compound.

6. The 3D imaging of a textile surface can allow us to detect defects which are not visible using conventional imaging.

7. Solvents have the ability to hold only a certain amount of solute, and then they become saturated.

8. One of the classic computer science problems is determining where data should be stored for optimal reading and writing.

9. In data-entry environments, clerks sit for hours in front of computer screens entering data.

10. Pesticides are applied directly to sheep to reduce parasitic infestation, and these residues are released into wool-processing wastewater during preparation.

二、汉译英

1. 如果我们想要获得合适的语言保真度，就需要 300 ~ 3300 赫兹的频谱。

2. 虽然常温状态下的水银在空气中是稳定的，但若受热便会与氧化合。

3. 有些蝙蝠，借助于接收它们自己发出的尖叫声的回声，就能探测出障碍物的位置并且避开这些障碍物。

4. 罗斯海的海水营养丰富，是南极海域最多产的部分，大量浮游生物和磷虾在此繁殖，因此也成了许多鱼类、海豹、企鹅和鲸鱼的家园。

5. 当混合剂燃烧时，由灼热的膨胀燃气所产生的推力的大小可以控制，方向也可以调整,从而推动火箭朝预定的方向飞行。

6. 较长的研制周期迫使人们主要根据对未来的预测来研制设备,而因涉及的高昂费用往往会推迟设备的投产时间。

7. 导线的直径和长度不是影响电阻的唯一因素。

8. 热固系统的强度高于冷固系统,且更抗老化。

9. 完成勘察工作之后,登上月球的两名宇航员便重新回到舱内。

10. 与卫星通信系统不同,标准高频发射机和接收机可以做到价格低廉、质量轻、体积小,并且只需要很小的功率即可工作。

第六章
拆译与合译

长难句是科技翻译的一大难点，主要原因在于英语和汉语有着迥异的表达习惯。诚如程洪珍指出，汉语重意合，属语义型语言，句子松散、简短；英语重形合，属形态性语言，短语和从句均可以充当句子的主次要成分，还有从句中嵌套从句的用法，从而形成了句子结构复杂、形式冗长的特点。拆译法和合译法正是解决该难点的利器。拆译法，也称分译法，指将原文中的单词或短语分离出来，改译为独立的句子，或将包含诸多分句的长句拆分成若干短语、分句或句子。合译法指将原文的各种复杂成分压缩合并，使译文更加紧凑、通俗易懂。针对英语中长句较多而汉语中短句较多的差异，英译汉时通常需要采取拆译法，而汉译英时则多使用合译法。

第一节　单词的拆译与合译

一般情况下，在英语和汉语中能够找到词义完全一致的单词，但也存在一些语义不完全对应甚至完全不对应的词。在处理一对多对应的词时，往往要采取拆译法；而处理多对一对应的词时，则采取合译法。此外，当翻译后某个单词的词义无法很好地融合在句中，或不符合目的语的表达习惯时，也需要采取拆译法将其单独翻译成分句。

副词/形容词的拆译与合译

(1) Not surprisingly, those who had eight hours of sleep hardly had any attention lapses and no cognitive declines over the 14 days of the study.

译文一　毫不意外地，在为期 14 天的研究中，每天睡 8 小时的人几乎没走过神，也没有出现认知能力下降的问题。

译文二　在为期 14 天的研究中，每天睡 8 小时的人几乎没走过神，也未出现认知能力下降的问题，这并不令人意外。

分析　原文中，副词短语“not surprisingly”作状语，位于句首修饰整个句子。译文一将其直译成副词，显得生硬刻板；译文二将其拆成一个独立的简单句“这并不令人意外”,并增译指示代词“这”,拉近与前文的语义联系。

(2) In the face of Internet-wide virus attacks, perhaps even more daunting is the realization that we will depend in larger and larger measure on the network's functioning reliably.

译文一 面对互联网范围内的病毒攻击,也许更令人生畏的是认识到我们将在越来越大的程度上依赖于网络的可靠运行。

译文二 面对互联网范围内的病毒攻击,我们会越来越依赖于网络的可靠运行,这更令人胆寒。

分析 原文是一个倒装句,主语中包含同位语从句显得较长,为了避免头重脚轻,将用来说明主语性状的表语“daunting”前置。译文一将对应的汉语副词“令人胆寒”同样放在了主语位置,并将后面的成分按原文顺序翻译,不符合汉语语序。译文二则将该形容词单独译为一个简单句并后置,整个句子先描述事实,再发表评论,符合汉语语序。

(3) 显然,润滑后的轴承,比未润滑的轴承容易转动。

译文 It is evident that a well lubricated bearing turns more easily than a dry one.

分析 原文中,“显然”作评论性状语,后面加逗号以便于区分出主语。译文将句子合译成了一个由形式主语“it”引导的主语从句,形容词“evident”作表语来评价真正的主语(润滑后的轴承)。

二 动词的拆译与合译

(1) Gasoline prices are not only affected by supply and demand but also by politics.

译文一 石油价格被供求关系和政治影响。

译文二 石油价格不仅受到供求关系的影响,还受到政治动荡的影响。

分析 原文中,关联词组“not only…but also”用于连接谓语动词,且由于谓语动词重复,“but also”后省略了动词“affected”。译文一并未补充已经省略的动词,而是直接将两个宾语合在一起。译文二则考虑到原文这一被动结构有两个施事,即“供求关系”和“政治动荡”,将谓语动词拆开翻译,使得两个分句间的并列关系更为明确,着重强调后者。

(2) Well-informed schoolchildren also know that the mass extinction at the end of the Cretaceous period was neither unique nor the biggest.

译文一 一些博学的小学生还知道,白垩纪末期的物种大灭绝不是唯一或最大。

译文二 见识多的学童还知道,白垩纪末的物种大灭绝既非史上唯一亦非最大的一次。

分析 原文是主系表结构,谓语动词为“be”,由表示否定的“neither…nor”连接两个并列成分。译文一按照原文结构将“be”动词直接译出,两个形容词表语并列,显得生硬。译文二充分考虑到动词后的两个表语以及汉语的关联词惯用表达“既非……亦非”,其处置更妥当,行文更紧凑。

(3) According to accounts of the experiments, women workers' hourly output rose when lighting was increased, but also it was dimmed.

译文一 根据实验的描述,当照明增加时,女工的小时产量增加,但也会变暗。

译文二 实验报告表明,当照明灯变亮时,女工们每小时的产出会增加,而当照明灯变暗时,她们每小时的产出仍会增加。

分析 原文由一个主句和两个时间状语从句组成,主要谓语动词是“rise”。两个时间状语从句间形成对比,分别描述“lighting”明和暗的情况。译文一将三个动词直接译出,堆砌在一起,无法看出其中的逻辑关系。译文二将谓语动词“增加”拆译,分别放入两个主句中,与两个时间状语从句一一对应,清晰明了地表达出照明灯的亮度与女工产出效率之间的关系,强调了“无论灯亮灯暗,产量都会增加”这一事实。

(4)大多数物种完全无法应对噪声对它们的影响,也无法适应拥挤的城市栖息地,最终只能逃到乡下去。

译文一 Most species are completely unable to cope with the impact of noise, nor to adapt to crowded urban habitats, and finally have to head to the country.

译文二 Most species simply can't handle the noise and cramped habitats of the city, and head to the country.

分析 原文有三个谓语动词,其中前两个动词“应对”和“适应”均是否定形式,在语义上相近,在结构上也相似,如果像译文一那样直接翻译,必定导致动词多用、句子相应变长。译文二则将两个动词合译成了一个,即“handle”,同时与跟随其后的两个宾语形成正常搭配,使得句子内容和结构简洁明晰,一目了然。

(5)在美国,每年有约1000万吨的纸被丢进了垃圾堆或倾倒在焚化炉中,这不仅造成了大量的纸张浪费,还增加了大气中二氧化碳的含量。

译文 Every year, about 10 million tons of paper winds up in American landfills and incinerators, which is not only wasteful but adds CO_2 to the atmosphere.

分析 原文中,动词“丢”和“倾倒”分别搭配“垃圾桶”和“焚化炉”,这是因为两个名词所代表的物体有大小之分,使用的群体也不尽相同。通常来说,普通人都可以使用垃圾桶,此时的动作是“丢”,即随手一扔;而焚化炉往往需要专门的工作人员来操作,比如将垃圾车上的垃圾从上往下倾倒。因此有了两个不同的动宾搭配。而在译文中,这两个语义相近的动词合译成了“winds up”。

第二节 短语的拆译与合译

短语是比单词更复杂的语言单位,因此也包含了更多的语义信息,需要通过拆译或合译的方法,将源语中的短语重新组合,按照目的语表达方式与语言习惯形成句子。

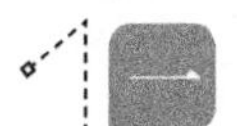

名词短语的拆译与合译

在科技英语中,名词短语充当主语时,可以跟随定语从句、同位语从句、分词短语等成分,还可以通过介词"of"连接至其他的名词短语。这样一来,名词短语不仅可以表示人、事、物、地点或其他抽象概念,还可以表达动作或状态,甚至是完整的句意。因此,可以将这些名词短语拆译成一个或多个分句,有助于完整表达各层含义。

(1) This hope of "early discovery" of lung cancer followed by surgical cure, which currently seems to be the most effective form of therapy, is often thwarted by diverse biology behaviors in the rate and direction of growth of cancer.

译文一 这种"早期发现"癌症,然后进行外科治疗的希望,可能是目前最有效的治疗办法,但经常因癌症生长速度和方向等生物学特征不相同而破灭。

译文二 早期发现肺癌,随即接受外科治疗,似乎是目前最有效的治疗办法。可是,由于肺癌生长速度和生长方向等生物特征差异很大,早期发现的希望常常化为泡影。

分析 原文的主语由两个"of"引导的介宾结构、一个分词短语和一个定语从句组成,长度较长且在汉语里没有对应的语法结构,如果像译文一那样直译,则语句不通顺。译文二将分词短语前的名词结构拆开,一个译成动宾结构放到句首(早期发现肺癌),另一个译成名词,充当分句的主语(早期发现的希望)。

(2) This lack of unique coloration on the sides of their bodies means that researchers can't usually tell if one puma crosses a camera trap five times, or if five individual animals pass by.

译文一 这种身体两侧独特的颜色的缺乏意味着研究人员通常无法区分是一只美洲狮经过抓拍相机五次,还是五只不同的美洲狮经过了抓拍相机。

译文二 它们身体两侧缺乏独特的颜色,这意味着研究人员通常无从断定是一只美洲狮经过抓拍相机五次,还是五只经过抓拍相机。

分析 原文的主语由一个"of"引导的介宾结构和一个介词短语组成,译文一将两个修饰成分均译成了带"的"字的形容词短语,放在名词前,由于句子较长显得更加生硬。译文二考虑到"lack"在汉英两种语言中都有动词、名词两个词性,将"lack"所在的名词短语拆分成了主谓宾结构的完整句子:名词"lack"转换成动词"缺乏"作谓语,名词短语"unique coloration"作宾语,介词短语"on the sides of bodies"作状语,代名词"their"转换成对应的代词"它们"作主语。

(3) Normally a blocked windpipe cuts off the blood's supply of oxygen, leading to brain damage and death.

译文一 正常情况下,堵塞的气管会切断血液的氧气供应,导致大脑损伤和死亡。

译文二 正常情况下,血管的阻塞会切断氧气供应,导致大脑损伤坏死。

分析 原文中,主语为冠词+分词+名词构成的名词短语"a blocked windpipe"。由于该部分较短,可以像译文一那样直接翻译。但译文二将名词短语拆译成单独的句子,并增译了假设关系连词"如果",使得前后两个分句之间的逻辑关系一目了然。

(4)他研究的是分叉环节动物,即像蚯蚓这样的动物,从幼虫阶段就长出了两个头,或者自然长出了两条尾巴。

译文 He looks at bifurcated annelids, meaning things like earthworms that have come out of their larval stage with two heads, or spontaneously sprouted two tails.

分析 原文是一个总分复句,名词短语"像蚯蚓这样的动物"及其后面的两个分句是对"分叉环节动物"这一专业术语的详细解释。译文将这种动物的两大特征合译成并列谓语,由"that"引导作定语从句,来修饰名词短语"things like earthworms"。

二 分词短语的拆译与合译

英语中的分词短语由现在分词或过去分词加名词或介词短语构成,可以理解为对从句的简化。一方面,在时间、原因、目的、条件等状语从句中,当从句的主语与主句的主语相同时,从句可简化为分词短语。另一方面,在定语从句中,关系代词用作主语时,从句可简化为分词短语。由于汉语中不存在分词短语,翻译时需要采取拆译或合译的方法来处理。

(1) The tape containing the information required to produce the part can be stored, reused or modified when required.

译文一 储存录制生产零件数据的磁带可以被保存、重复使用或修改,需要时。

译文二 储存录制生产零件数据的磁带可以保存起来,重复使用。必要时,还可以修改数据。

分析 原文中,分词短语可以改写成时间状语从句"when it is required",修饰的是连词"or"后的动词,即"modify"。译文一直接将三个并列的谓语动词同时翻译,丢失了原文的重要信息——时间状语"when required"与动词"modify"之间的关系。

译文二将分词短语及其前面的动词拆译成一个句子，突出信息焦点，并增译副词“还”，表明“修改数据”与上文的“保存”“重复使用”是递进关系。

(2) An antibody-based blood test can detect a particular form of Tau protein called brain-derived Tau, which is specific to Alzheimer's disease.

译文一　基于抗体的血液测试可以检测出一种特殊形式的被称作脑源性 Tau 蛋白的阿尔茨海默病患者特有的 Tau 蛋白。

译文二　基于抗体的血液测试可以检测出一种特殊形式的 Tau 蛋白，人称脑源性 Tau 蛋白。这种蛋白是阿尔茨海默病患者特有的。

(3) A number of universities and companies have been removing online data sets containing thousands of photographs of faces used to improve facial-recognition algorithms.

译文一　一些大学和公司已经删除了包含数千张用于改进人脸识别算法的人脸照片的在线数据集。

译文二　一些大学和公司一直在删除包含数千张人脸照片的在线数据集，这些数据集用于改进人脸识别算法。

分析　在以上两例中，“called”和“used”的分词短语均为后置定语，分别修饰“Tau protein”和“photographs of faces”。译文一直接译成带“的”字的修饰性短语，置于名词前，句子冗长，信息不突出。译文二将分词短语拆译成小句，简洁有力。其中例(3)还增译了主语“这些数据集”，与上一个分句的定语相呼应。

三　介词短语的拆译与合译

“with”和“without”是英语中十分常见的介词，其用法却较为复杂。除了介词常见的与名词连用的用法外，这两个介词还可以接宾语和宾语补足语，构成复合结构，在句中作状语，表示条件、伴随、方式等。译成汉语时，可以适当拆分。

(1) Taking a page from the art-restoration handbook, scientists sampled a variety of light sources to see if any could be used to strip the ink from laser-printed documents without damaging or discoloring the paper.

译文　科学家从艺术修复手册上取下了一张纸，对多种光源进行了取样，看能否去除激光打印纸上的油墨，而不让纸张损坏或褪色。

分析　原文中，“without”引导的介词短语在句中作条件状语，表明合格光源的标准是既可以去除油墨，又不会对纸张造成负面影响。译文按照动词将原文拆成四个分句，其中介词短语译成了主谓结构的简单句，并增译连词“而”，表明本句和上一个分句都是衡量光源合格与否的标准，缺一不可。

(2) Space Post Office adopts the operation mode featuring the combination of physical and

virtual one, with the former established at the post office of the Space City in Beijing and the virtual one in the spacecraft.

译文 太空邮局采取“虚实结合”的经营模式,实体邮局设在北京航天城邮局,虚拟邮局设置在空间飞行器内。

分析 原文中,“with”引导的介词短语是省略式,可以理解为两个并列的、由“and”连接的“with + 宾语 + 过去分词”结构,是对上文“虚实结合”经营模式的解释说明。译文补充了省略的部分,将该介词短语拆译成两个主谓结构的并列分句。

(3)已经观察到60多种蠕虫存在分叉现象,这种现象遍及环节动物族谱。

译文一 Over 60 species of worms have been observed to have bifurcations. This phenomenon is common in the annelid family tree.

译文二 Bifurcation in worms has been observed in over 60 species of worms across the annelid family tree.

分析 原文由两个主谓宾结构的分句组成,第二个分句的主语即是第一个分句描述的事件。相比于译文一直译成两个句子,译文二选择“bifurcation”一词为主语,可以搭配原文中两个谓语,因此第二个分句中的介词短语直接并入主句,表示范围,修饰谓语动词“observe”,更加简洁有力,符合英语表达习惯。

第三节 句子的拆译与合译

句子是比词、短语更大的翻译单位,更能体现英汉两种语言的差异,翻译起来也更为困难。从句法结构上,可以将句子分为简单句和复合句。

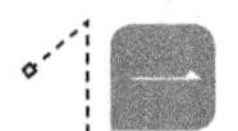

一 简单句的拆译与合译

简单句只有一个主谓结构,是最短的句子单位。在英语中,一个句子只能有一个主语和一个谓语动词,当描述不同主语的动作时,即使两个主语之间存在联系也需要另起一句;而汉语中,一个句子内部不仅可以频换主语,还可以连续出现多个动词。因此,英译汉时可以将两个相邻的简单句合成一个句子,汉译英时可以将一个简单句拆分成两个存在一定逻辑关系的句子。

(1) When the asteroid hit Earth, it sent vaporized rock particles above the atmosphere. There, the particles would have formed into grains of sand and then re-entered the atmosphere.

译文一 当行星撞击地球时,它将蒸发的岩石颗粒送入大气层。在那里,这些颗粒会形成沙粒,然后重新进入大气层。

译文二 当行星撞击地球时,蒸发岩粒子会喷到大气层上,在大气层以外的地方组成无数沙粒后回落到大气层。

分析 原文由两个独立的简单句组成,其中“vaporized rock particles”既是第一个句子的宾语,又是第二个句子的主语。译文一的直译使得句子结构松散,句子与句子之间联系较少。译文二将两个简单句合译成一个并列句,由“蒸发岩粒子”作“喷到大气层上”“组成无数沙粒”和“回落到大气层”这三个动作的执行者,既是施事也是主语。

(2) Long haulers can have lingering symptoms including fatigue, body aches, inability to exercise, headache, and difficulty sleeping. Some of these problems may be due to damage to lungs, heart, kidneys, or other organs.

译文 迁延不愈者可能出现疲劳、身体疼痛、无法运动、头痛、睡眠困难等后遗症,这可能是由肺损伤、心脏损伤、肾脏损伤等器官损伤造成的。

分析 从介词短语“due to”可以看出,原文中两个简单句之间必存因果关系,第二句与第一句首尾相呼应,均表示事件的结果。译文将其合译成一句,通过“是由……造成的”这一结构来表示因果关系。

(3) Stilbella aciculosa 真菌长有芽孢,也分泌超氧化物(一种高活性的氧)。

译文 As Stilbella aciculosa makes spores, it also produces superoxide. That's a highly reactive kind of oxygen.

分析 原文中,括号里的内容是对宾语“超氧化物”的补充说明。译文将其单独译成一句,并通过指示代词“that”与上文建立紧密联系。

(4) 强电流能在一瞬间摧毁变压器和断路器,同时还能通过弱化金属腐蚀油气管线。

译文 In an instant, the surges destroy transformers and overwhelm circuit breakers. They also corrode pipelines by weakening away the metal.

分析 原文中有两个谓语动词,即“摧毁”和“腐蚀”。由于“摧毁”与两个宾语搭配,在译为英语之后,两个宾语需要搭配不同的动词(“destroy”和“overwhelm”),这样一来,译文就存在三个动词和一个现在分词。因此,译文将简单句拆分成两句,分别描述强电流的两种功能,第二句中人称代词“they”用作主语,避免重复,又对上文起到回指作用。

二 复合句的拆译与合译

复合句是由一个主句加上一个或多个从句构成的句子,后者对前者起到修饰作用,两者通过从属连词引导。常见的复合句主要包含定语从句和状语从句两大类型。如果原文结构

较为简单,修饰成分较少,且译成汉语后的从句较短,那么可以直接合译到主句中,如:

(1) The real danger is the height of hypercanes, which would have been 40 miles.

译文 真正的危险在于时速可能高达40英里的超级飓风。

但在大多数情况下,译文更倾向于遵循汉语短句多的习惯,将从句拆译,并调整语序,使之单独成为一个分句。

(2) The comet was believed to have come from the Oort Cloud, a vast sphere surrounding the solar system that is home to mysterious icy objects.

译文一 这颗彗星被认为来自奥尔特云,这是一个围绕太阳系的神秘的冰天体的家园的巨大球体。

译文二 据称该彗星来自奥尔特云,这片巨大区域位于太阳系外围,是神秘的冰冻天体的家园。

分析 原文中包含了一个"that"引导的限制性定语从句,长度较长且句中还有其他修饰成分,译文一把所有修饰成分置于名词前,"的"字堆叠,冗长不清,语义模糊。译文二则很好地采用了拆译法,将原文中的同位语"这片巨大区域"作为主语,两个修饰成分定语从句和分词短语作并列谓语,两个分句共同对上文起解释说明的作用。

(3) The speed of sound through water depends on the physical properties of that water, which are related to temperature.

译文一 声音在水中的传播速度取决于与温度有关的水的物理性质。

译文二 水中的声速取决于水的物理性质,而这些性质又与温度有关。

分析 原文中,"which"引导的非限制性定语从句与主句在逻辑上存在递进关系,即声音→水→温度,如果像译文一那样直译,不仅无法忠实传达这一逻辑关系,还会使读者对三者之间的关系产生疑惑。译文二则将其单独译成汉语小句,并添加连接词"而",使两个分句之间的递进关系显性化。

(4) In an effort to tame its air pollution, the city of Hong Kong has deployed a system that can sense when a high polluting vehicle drives by.

译文一 为了治理空气污染,香港市区已经部署了一种当高污染的车辆驶过时可以感知到的系统。

译文二 为治理空气污染,香港市区已经部署了一种感应系统。当高污染车辆驶过时,该系统即可感应到。

分析 原文除主句外,还包含一个定语从句和一个状语从句。译文一将两个从句翻译成前置定语修饰名词,使得句子冗长而表意不清。译文二将其单独拆分成一个完整的句子,时间状语从句前置,增加连词"当……时",同时增译定语从句的先行词"该系统"用作主语,使得句子更加通顺。

练习题

一、英译汉

1. A computer is a device which takes in a series of electrical impulses representing information, combines them, sorts them, analyses and compares the information with that stored in the computer.

2. More fire-favorable weather associated with declines in the Arctic Sea ice during summer can increase autumn wildfires over the western United States.

3. Recent discoveries that certain plastics can conduct electricity are arousing industrial and academic interest, because of the promise of new technology exploiting the high conductivity of these materials which is comparable to that of costly metals.

4. Made of ice and dust and emitting a greenish aura, the comet is estimated to have a diameter of around a kilometer.

5. Scientists have developed a blood test to diagnose Alzheimer's disease without the need for expensive brain imaging or a painful lumbar puncture.

6. A newly discovered comet could be visible to the naked eye as it shoots past Earth and the Sun in the coming weeks for the first time in 50,000 years.

7. Their segmented bodies, like an earthworm with rows of ringed compartments, help them easily regrow a new head or tail at the first sign of trouble.

8. Hypercanes brought water into the stratosphere and deteriorated the ozone, which would have killed off any creatures that couldn't find shelter until the ozone reformed.

9. The brick battery relies on the reddish pigment known as iron oxide, or rust, that gives red bricks their color.

10. They attached suction-cup sensors to the outside of the whale's head, which measured brainwaves that indicated the whale did reduce its hearing sensitivity in expectation of a clamor.

二、汉译英

1. 很显然的是，继续开发更多的节能技术将很有必要。

2. 卫星提供了海面的信息，科学家们已经发射了漂浮装置用来测量水面上一英里范围的情况。

3. 如今，科学家们已经开发出一种新技术，可以测量整个海洋盆地的温度变化。

4. 气候变暖加快了氮循环，使更多的氮气用作植物的肥料，但是也有大量的氮通过径流流失或以气态逸出从而离开土壤。

5. 随机数字可以用来加密信用卡和邮件，对现代计算系统极为重要。

6. 很多所谓的随机数字实际上不是随机的，而是由算法产生的“伪随机数”。

7. 要充分重视自然科学基础理论的研究,忽视这一点就不能掌握和应用世界上先进的科学技术成果,也不能妥善解决我国建设中遇到的重要问题。

8. 亚马孙河流域有世界上著名的复杂生态系统,中国科学家于2004年7月首次到该地区进行科学考察研究。

9. 上海磁悬浮,指上海磁悬浮列车专线,西起上海轨道交通2号线的龙阳路站,东至上海浦东国际机场,全长29.863千米。

10. 中国火星探测计划是中国首个火星探测计划,由中国国家航天局与俄罗斯联邦航天局合作,共同探索火星。

第七章 词性转换

“翻译是用最恰当、自然和对等语言从语义到文体再现源语的信息。”Nida从语言学的角度提出了著名的“动态对等”翻译理论，并将其运用于实践中。英语和汉语是不同语系的两种语言，前者属于印欧语系，后者则是汉藏语系。两者在语言结构上有所差异，在词性上也很难做到一一对应。科技文本属于信息型文本，其翻译重点在于达意，以传递信息为主。因此，目的语在形式上并不要求与源语一一对应。“内容比形式重要”是科技翻译的重要准则之一。英汉互译时，为了让有不同文化背景的读者获得相同的理解和阅读感受，灵活运用词性转换，可增加译文流畅感，避免生硬的表达，摆脱行文板滞，且有利于更加完整准确地传达原文的信息。

本章将讨论英汉互译中四种常见的词性转换情况，并对科技英语中名词化这一显著特征做简单介绍。

第一节　转译为动词

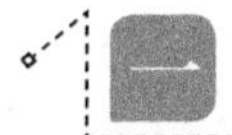

一　名词译为动词

(1) The second reason for the importance of materials in an energy program is that the development of new materials and the improvement of existing ones have a direct impact on the cost of energy.

译文一　在能源项目中，材料之所以重要的第二个原因是：新材料的开发与现有材料的改良对（降低）能源成本具有直接影响。

译文二　在能源项目中，材料之所以重要的第二个原因是：开发新材料与改良现有材料对（降低）能源成本具有直接影响。

分析　比较上面两个译文后发现，后者明显更加符合汉语表达。在英语中，这种由动词派生而来的名词有逻辑宾语，表现的形式是“名词＋介词＋名词”。在英译汉时，采用从前到后的顺译法，译成汉语中的“动宾结构”。

(2)This can explain why some people living in the modern cities commit suicide while they enjoy the real boons that advanced technology provide with.

译文一 这可以解释为什么一些现代城市人在享受先进技术带来的益处时,仍会采取自杀。

译文二 这或可解释一些生活在现代都市的人在享受先进技术提供的益处时,仍会自杀。

分析 上面两个译文的不同之处在于对“commit suicide”的翻译。英语中有一些含有主体名词的动词短语,在翻译成汉语时,直接将含有动作意义的主题名词转译为动词,如“give a suggestion”,转译为“suggest”即可。

(3)They were the prime movers in all the inventions of the stage.

译文一 他们是这个阶段所有的主要发明创造的推进者。

译文二 这个阶段所有的主要发明创造都是他们推进的。

分析 英语中一些以“-er”或“-or”结尾且含有较强动作意味,但不是表示职业或身份的名词,英译汉时通常将其译为动词。本例中,“movers”不是职业,只需译出“move”的动作意义,用汉语“是……的”这一强调句型来连接前置的宾语。

(4)这艘船容易倾翻是因为它的造型不合理。

译文一 This boat is prone to overturn because its shape is unreasonable.

译文二 This boat is prone to overturn because it is unreasonably shaped.

分析 比较上面两个译文,译文一满足了形式上与原文的对应,意思清楚,但是过于僵硬呆板。显然,译文二中“unreasonably shaped”更加符合科技英语的“副词 + 动词衍生而来的形容词”这一常见表达。

(5)纽约就是这样一座城市。它的街道或如蛇形,或呈环状,或如锯齿。

译文一 New York is such a city, whose streets are like snakes, circles and zigzags.

译文二 New York is such a city, whose streets snake, circle and zigzag.

分析 译文一在词性上与原文相对应,且遵从其行文方式,为可接受的译本。译文二将原文中的“蛇形”“环状”“锯齿”三个汉语名词均翻译成简洁有力的动词,使整个句子呈现出动态感,更容易吸引读者的眼球。

(6)这座工厂先进的生产线和高效的生产组织给我们留下了深刻的印象。

译文一 The factory made a strong impression on us that it has cutting-edge production lines and efficient organization of production.

译文二 The factory impressed us deeply with its cutting-edge production lines and efficient organization of production.

分析 比较两句译文,译文一使用了“made a strong impression on us that…”这一“弱

势动词＋名词＋that 引导的同位语从句”结构，使得整个句子显得冗长复杂。译文二将原文中的“留下了深刻的印象”译成动词短语，使得整个译句简洁流畅。

二 形容词译为动词

(1) There has been a diminishing requirement for physical effort, both in our occupations and our daily life, as a direct result of scientific and technological progress.

译文一 科学技术的进步直接导致职业活动和日常生活对体力的一种减少的需求。

译文二 科学技术的进步，直接导致职业活动和日常生活对体力的需求减少。

分析 比较两个译文，“减少的需求”这种说法在汉语中较为罕见。原文中的“diminishing”派生自动词“diminish”，译文二选择将其转译为带动词意味的“减少”，使整个译文更贴近汉语结构或表达。

(2) Encoding these parameters causes shorter chromosomes and, consequently, shorter computing time.

译文一 编码这些参数使染色体较短，从而让计算时间短些。

译文二 编码这些参数导致染色体变短，从而缩短计算时间。

分析 在英译汉时，英语形容词修饰名词的这类固定搭配并非可以完全对等地译为汉语中的对应结构。“consequently”一词表示前后两个小句带有因果关系，将第二个形容词“shorter”转译为动词“缩短”更能凸显这种前后小句紧密的联系。另外，该学科领域一般用“导致染色体变短/缩短”，译文二则是在第一个“shorter”前增译了一个动词。

(3) If a plant is placed on the condition devoid of oxygen, it very soon replenishes it.

译文 植物若处于缺氧状态，它很快就会补充氧气。

分析 英语中有一类表示动态行为的形容词具有较强的动词意义，如“scarce”“sure”等。它们往往出现在“be＋形容词＋of”这类结构中。在英译汉时，通常根据语境将这类形容词翻译为动词“匮乏”“肯定”。

三 介词译为动词

英语中有许多含有动作意义的介词，当它们在句子中作表语或状语时，英译汉时可转译为动词。而在汉译英时，一个句子往往有两个及以上的动词，斟酌将其中不处于主干的动词翻译为英语中的介词，可以使得结构更为紧密，条理更为清晰。

(1) Numerical simulations play a key role in the design of cutting-edge suspensions and in such simulations the road inputs can be regarded as stochastic processes.

译文一 数值仿真技术在先进悬架的设计方面起着非常重要的作用,在这种仿真技术里面,路面输入可以被视为一个随机过程。

译文二 在先进悬架的设计方面,数值仿真技术起着非常重要的作用,运用这种仿真技术我们可以把路面输入视为一个随机过程。

分析 "in"最常见的用法是表示持续时间的某个节点、方位指向或空间结构内部。而这里是表达抽象含义"在……里面",译文一将其直译并在结构上完全生硬地对应原文,造成译文不流畅,影响其可读性。原文所表达的意思是数值仿真技术在先进悬架的设计方面的重要应用功能,所以将"in"转译为汉语中的"运用"更为恰当。

(2) The promotion of this robot's final path is that the people in the underdeveloped areas will be eventually off poverty.

译文 推广这类机器人,目的是使不发达地区的人们最终摆脱贫困。

分析 原句中的"off"与"He's gone to sleep off a headache."中的"off"相同,都为"从某种状态脱离出来"。因此,在翻译成汉语的时候,需要将其转译为符合当前语境的含义,转换为汉语动宾搭配即"摆脱贫困"。

(3) Radio telescopes have been able to probe space beyond the range of ordinary optical telescopes.

译文 射电望远镜已能探测普通光学望远镜所不及的宇宙空间。

分析 "beyond the range of ordinary optical telescopes"属于英语中的介宾搭配,实际是具有动态含义的结构。翻译时需要把握整个句子,以达到最方便译文读者理解文本信息的要求。

(4) 在20世纪40年代,美国城市居民为了寻求新鲜的空气、充裕的活动场所和独家独院的小天地而大量迁往郊区。

译文一 In the 1940s, many urban Americans moved to the suburbs to search fresh air, elbow room, and privacy.

译文二 In the 1940s, many urban Americans moved to the suburbs in search of fresh air, elbow room, and privacy.

分析 整个句子主要的谓语动作是"迁往……",这能在译文中得到对应的动词短语"moved to..."。译文一使用动词"不定式 to + 动词原形"这一结构表目的,然而前文已有一个表示移动方向或地点的"to",如此重复用词影响了科技英语文本的准确性和严谨性。而在译文二中"寻求"这一动词搭配三个名词短语结构,转译为表示目的的介词短语"in search of...",更加符合英语语感。

第二节　转译为名词

动词译为名词

(1) This technology forces the engine to operate at lower combustion temperatures, reducing the quantity of nitrogen-oxide produced.

译文一　这一技术迫使发动机能在较低燃烧温度下运转,减少了氮氧化物的产生。

译文二　这一技术迫使发动机的运转能在较低燃烧温度下进行,减少了氮氧化物的产生。

分析　原句强调的是"发动机能够在较低燃烧温度下运转"这一动态过程。而译文一前半句的主语是"发动机",不能突显运转过程。译文二将英语动词"operate"翻译成汉语名词"运转"并将其作为主语,使其得到强调,是后半句的直接原因。

(2) This system will build on the knowledge and experience gained in developing the existing suite of software products.

译文一　这个系统将建立在开发现有的成套软件产品中所获取的知识和经验上。

译文二　该系统的基础源于开发现有的成套软件产品中所获取的知识和经验。

分析　原文动词短语"build on"的受事"knowledge and experience"的定语较长。如果将其翻译成为汉语动词"建立"(译文一),则必然出现"建立在……上"这一封闭式句型,致使成分过长,可读性较差。译文二采取了将英语动词翻译成为汉语名词的策略,采用开放式结构,既准确无误地表达了原文的信息,又符合汉语的表达习惯。

(3) Large language models, trained on vast amounts of data to predict deleted words, have an uncanny ability to mimic the patterns of real language and say things that humans might.

译文一　以海量数据为基础训练,预测删除词的大语言模型具有一种不可思议的能力,它可以模仿人类语言的模式,说出人类可能说的话。

译文二　以海量数据为基础加以训练,预测删除词的大语言模型具有一种不可思议的能力,它可以模仿人类语言的模式,说出人类可能说的话。

分析　原句中的"trained"为动词的过去分词形式表被动,表示该"大语言模型"是在通过训练后得以运用。译文一将"基础训练"视为整体,因此存在歧义;若不视为整体,则导致动词"训练"之后缺少宾语。而在汉译时应当考虑汉语使用主动语态更多这一实际,增加形式动词"加以"与转译为汉语名词的"训练"搭配,使句子更符合汉语表达。

(4)他的最重要的革新是在高年级学生中采用启发式教学法。

译文一 His most important innovation was to use the heuristic method of teaching for advanced students.

译文二 His most important innovation was the introduction of the heuristic method of teaching for advanced students.

分析 译文一采取了直译的方式,这样使得前面出现系动词"was"后再次使用了动词"use"。而译文二将"采用"一词转译为英语名词"introduction",不仅减少了句子中动词的使用频率,让名词主语和名词宾语之间的结构更加平衡,还让"启发式教学法"这一句子焦点得以凸显。

二 形容词译为名词

(1) Any predictive model will generate false positives, in which innocent people are flagged for investigation.

译文一 任何预测性的模型都会出现误报,无辜的人被标记为调查对象。

译文二 任何预测模型都会出现误报,将无辜的人标记为调查对象。

分析 原句中的"model"由形容词"predictive"修饰,限定了这一模型的功能性质。而在汉语中,名词可以修饰名词,其作用和形容词一样,往往是说明被修饰名词的材料、用途、时间、地点、内容、类别等。

(2) A bridge engineer must have three points in mind while working on a bridge project: ① creative and aesthetic; ② analytical; ③ technical and practical.

译文一 桥梁工程师在设计桥梁时必须牢记三点:①是创新的和美观的;②是可分析的;③有技术水平的和实用的。

译文二 桥梁工程师在设计桥梁时必须牢记三点:①创新性和美观性;②可分析性;③技术水平和实用性。

分析 对比两个译文,译文二采用词性转换策略后,将原文中的五个英语形容词全部转译为汉语名词,这使得译文简洁有力,结构严谨,而译文一采用直译方法,虽保留了原文词性,译文却显得生硬,有碍读者阅读。

(3)有时雄性成分到达胚珠在客观上是不可能的,例如雌蕊太长以至于花粉管无法到达子房的植物就是这种情况。

译文一 Sometimes it is physically impossible for the male component to reach the ovule, as is the case with plants where the pistil is so long that the pollen tube cannot reach the ovary.

译文二 There must sometimes be a physical impossibility in the male element reaching the ovule, as would be the case with a plant having a pistil too long for the pollen-tubes to reach the ovarium.

分析 译文二在将原句翻译成为英语时采用了"There + be"句型,这使得后面的句子成分的词性受到了一定的限制,译文二中的"a physical impossibility"为专业术语,相比译文一,这增强了科技英语文本的专业客观性。

三 副词译为名词

(1) All this suggests that copper could be drilled for commercially in the same way that oil is except that the boreholes involved would be considerably deeper.

译文一 所有这些都表明铜可以像石油一样在商业上被开采——除了钻孔要深得多之外。

译文二 所有这些都表明铜可以像石油一样用于商业开采——除了钻孔要深得多之外。

分析 原句中的"be drilled for commercially"是表示开采的目的是发展商业,在翻译"commercially"时,将其翻译成为名词也可修饰名词"开采"。译文一完全照应原文的被动语态,使得译文生硬不流畅,加大了读者理解难度。

(2) Whereas DNA samples from a living animal can run to several hundreds of thousands of letters, the timeworn mammoth samples yielded strands mere dozens of letters long. This is close to the limit of what is scientifically usable.

译文一 活体动物的 DNA 样本可以有几十万个字母,而年代久远的猛犸象样本中的 DNA 样本只有几十个字母长,这接近科学上可用的极限。

译文二 活体动物的 DNA 样本可能有几十万个字母,而年代久远的猛犸象样本中的 DNA 样本只有几十个字母长,这接近科学使用极限。

分析 "scientifically"由名词"science"派生而来,译文一按字面意思将"scientifically usable"生硬地处理为"科学上可用的"。对比两个译文,译文二考虑到整个句子内部成分的协调,将其还原为名词性结构能让译文更加简洁,更满足科技英语文体特征的要求。

(3) Precisely simulating all but the simplest chemical reactions is mathematically intractable for any non-quantum computer, no matter how huge.

译文一 精确模拟哪怕是最简单的化学反应,对任何一台非量子计算机而言,无论是多大的非量子计算机,都是在数学上相当棘手的。

译文二 哪怕是精确模拟最简单的化学反应,所需处理的数学难题对任何一台非量子计算机而言,无论是多大的非量子计算机,都是相当棘手的。

分析 首先,译文一与原文句子结构对应的翻译让整个译文生硬不流畅。其次,译文一未处理"mathematically intractable"这一结构,直译为"在数学上相当棘手的",不符合汉语表达习惯。对比两个译文可发现,译文二调整了句子结构,逻辑合理,也将"mathematically intractable"拆开处理,"都是相当棘手的"总结了原文表达处理"数学难题"有难度的含义。

(4)增加这种装置将保证工件装卸方便。

译文一 The addition of this device will ensure that the loading and unloading of workpieces is easy.

译文二 The added device will ensure accessibility for part loading and unloading.

分析 在英译时,译文二将"装卸"一词拆分为"loading"和"unloading",并使用连词"and"连接。将副词"方便"译成英语名词"accessibility",并使用介词"for"提示作用对象,使句子结构严谨,符合科技英语文本特征。译文一使用的"the addition of..."这一名词短语和"that"引导的宾语从句使得整个译文结构冗长,不利于译文读者理解,降低了该科技文本的可读性。

四 代词译为名词

所谓英语代词转译为汉语名词,就是在翻译时将句子中所指代的名词还原。

(1)The radioactivity of the new element is several million times stronger than that of uranium.

译文一 这种新元素的放射性比铀的那个强几百万倍。

译文二 这种新元素的放射性比铀的放射性强几百万倍。

分析 英语中常使用"that"指代前文已出现并拿来做比较的对象,因此可以联系前文将其还原为语境中的名词。译文一直译了"that"的指代意义,造成译文不够严谨,而译文二将其指代的具体内容翻译出来,使得译文更具逻辑性,传达的信息是完整的。

(2)One would fall all the way down to the center of the earth without gravity.

译文一 如果没有重力,一个人会一直跌落到地球中心。

译文二 如果没有重力,人会一直跌落到地心。

分析 "one"在英语中指代对象广泛,翻译时需要根据其在句子中的实际情况灵活处理。此句中假设的是人类失去重力后将会发生的情况,对比两个译文,前者使用"量词+名词",后者则是直接使用名词,泛指整个人类,描述了普遍的情况,逻辑性更强。

第三节　转译为形容词

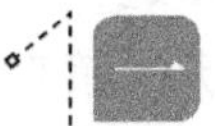

名词译为形容词

(1) Friction may be a nuisance to designers of some machines, but it is a necessity for everyday life.

译文一　摩擦力也许是使某些机械设计人员感到伤脑筋的问题，但它又是日常生活中的一个必需品。

译文二　摩擦力也许是使某些机械设计人员感到伤脑筋的问题，但它又是日常生活中所必不可少的。

分析　英语中某些名词前加不定冠词作表语“a/an”具有形容词功能，如“success”“stranger”。在英译汉时，这类名词通常翻译成形容词（如译文二）。摩擦力是抽象存在于日常生活之中的，译文一使用“量词＋名词”来形容它是“一个必需品”不符合一般认知，较为怪异。

(2) Single crystals of high perfection are an absolute necessity for the fabrication of integrated circuits.

译文一　有高度完美性的单晶对制造集成电路来讲是绝对必要的。

译文二　高度完整的单晶对制造集成电路来讲是绝对必要的。

分析　像“perfection”“difficulty”“importance”这类由形容词派生的、通常用来描述事物的属性或特征的名词，在句子中与介词搭配，充当表语和定语，一般译为汉语形容词。

(3) 裂变碎片在裂变过程中的放射性很强。

译文一　The radioactivity of fission fragments in fission process are very strong.

译文二　The fission fragments in fission process are very radioactive.

分析　对比两个译文，译文一没有采用词性转换策略处理“放射性”这一名词，其主语“radioactivity”的修饰语又过长，这使得整个译文头重脚轻。而译文二选择了同原文一致的主语——“裂变碎片”（the fission fragments），同时，将“放射性”转译为英语形容词“radioactive”，整个句子逻辑清晰，结构严谨，增加了译文的可读性。

二　动词译为形容词

(1) Noncryogenic methods of air separation contain some of the same unit processes such as

compression and clean-up, but differ in the fundamental technology to separate air into its components.

译文 非低温空气分离方法使用一些相同的单元过程,如压缩和清洁,但在把空气分解成它的组成成分的基本技术上是不同的。

分析 原句中"...some of the same unit..."强调的是相同之处,"differ"则是进一步说明该方法在把空气分解成它的组成成分时又是不同的,将其翻译成为汉语形容词能够前后对照,使句子结构对称,符合汉语语感。

(2)世上只有海水取之不尽,用之不竭。水资源的确能为人类生存发展提供无限动力。

译文 Only the sea water is inexhaustible. It is indeed the water that act as an impetus to the existence and advancement of human beings.

分析 "取之不尽,用之不竭"在这里描写海水数量多这一状态,在原文中充当谓语。在英译时需要寻找英语中描写状态的语言单位,即形容词。"inexhaustible"意为"用不完的,无穷无尽的",是符合这一语境的含义。同时需要注意的是,原文中用一组成语来强调这一状态,而英译时为了使句子更为简洁,只取了最为核心的含义。

(3)研究人员使用氮离子(氮的一种形式)开展了这项测试。他们发现,与正水分子相比,负水分子化学反应速度提高了25%左右。

译文一 The researchers used ultracold diazenylium ions (a form of nitrogen) for this test. They found that the reaction speed of para-water is improved about 25% with the diazenylium comparing to ortho-water.

译文二 The researchers used ultracold diazenylium ions (a form of nitrogen) for this test. They found that para-water reacted about 25% faster with the diazenylium than ortho-water.

分析 对比两个译文,译文一将"提高"这一动作英译成被动语态以满足句子的语义逻辑,译文二则将其转译为英语形容词且使用比较级,这使得译文结构简洁,前后逻辑紧密,更加符合科技英语文本的特征。

三 副词译为形容词

英语中的某些动词在翻译为汉语时会转译为汉语里的名词,因此英语中原来修饰动词的副词也随之转译成汉语中相应的形容词。

(1)Plants and pollen tend to be negatively charged, and bees are positively charged.

译文一 植物和花粉通常被带上负电荷,而蜜蜂通常被带上正电荷。

译文二 植物和花粉通常带负电荷,而蜜蜂通常带正电荷。

分析 译文一将原文的被动语态翻译了出来,然而汉语中可以用主动态来表示被动义。"be negatively charged"若按照成分一一对应翻译的话则为"被负电性地充电",完全违背汉语表达习惯。因此,将原文中的动词汉译成汉语名词,原句中的副词也随之译为汉语形容词,并增译动词"带",使得句子更为顺畅。

(2) A growing body of evidence shows that nuclear power is the most environmentally friendly energy of the future.

译文一 越来越多的证据表明,核能是未来环境上最友好的能源。

译文二 越来越多的证据表明,核能是未来最为环保的能源。

分析 译文将原句中的副词"environmentally"和形容词"friendly"杂糅后翻译成为"环保的",使得译文简洁有力。且这两个单词已经成为固定搭配,例如"environmentally friendly washing powder"(环保洗衣粉)。

(3) The technology is developing rapidly and one can easily imagine entire aircraft assembled out of printed parts.

译文一 技术发展迅速,人们可以很容易就能想象出整架飞机是由印刷部件组装而成。

译文二 随着技术的迅速发展,人们很容易就能想象出整架飞机是由印刷部件组装而成。

分析 译文二用"随着……"这一普遍接受的欧化汉语结构来表达原文"and"所连接的两个小句间内在的因果关系。具有动词含义的"developing"在汉译时转译为汉语中的名词"发展",因此修饰它的副词"rapidly"也随之转换成形容词。

(4) 在科学研究、设计和经济计算方面广泛地应用电子计算机可以使人们从繁重的计算工作中解放出来。

译文 The wide application of the electronic computer in scientific work, in designing and in economic calculations will free man from the labor of complicated computations.

分析 在翻译"使人们从……解放出来"这一结构时,译文选择情态动词"will",并将动词"解放"作为句子的核心动词,这样一来谓语前面就需要一个名词性结构作主语。"应用"一词转译为名词后,修饰它的副词"广泛地"也随之转译为形容词。

第四节 转译为副词

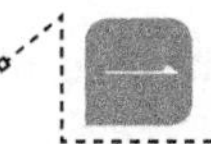

一 形容词转译为副词

(1) Fast Ethernet limits the distance between a computer and a hub to only 100 meters, making careful network planning a necessity.

译文一 快速以太网把计算机与集线器之间的距离限制在仅 100 米,这让细心的网络设计成为一个必要。

译文二 快速以太网把计算机与集线器之间的距离限制在仅 100 米,这就必须细心地设计网络。

分析 形容词修饰的名词在汉译时转译为动词,该形容词也将随之转译为汉语副词以满足搭配需求。"planning"转译为汉语动词"设计"后,修饰它的形容词"careful"也应翻译为副词"细心地",因此译文二比译文一更合理。

(2) Hence there is a pressing need to make realistic assessments of probable demand for the whole range of industry resources taking into account past trends, future population structure and the various factors influencing demand.

译文一 因此,对整个产业资源的需求量进行实际评估时,有一个急切的需求去考虑过往的需求走势、未来的人口结构和影响需求的各种因素。

译文二 因此,对整个产业资源的需求量作出实际评估时,急需考虑过往的需求走势、未来的人口结构和影响需求的各种因素。

分析 形容词"pressing"修饰名词"need",后者在译文中转译为汉语动词"需(要)",前者也随之转译为汉语副词"急"。相比于译文一使用"量词 + 名词"直译原文,译文二在使用词性转换策略后的译文更加符合汉语表达,从而更具可读性。

(3) 地球内的岩浆可能含有大量的气体和蒸汽。

译文一 The magma within the earth may have a lot of gases and steam.

译文二 The magma within the earth may be heavily charged with gases and steam.

分析 译文一和译文二分别使用了主动语态和被动语态。"气体和蒸汽"是一种弥漫式分布在空气中的物质,使用被动语态不仅能强调是它们对空间的控制,还能凸显科技英语文本的客观性。译文一将原文"含有"直接翻译为"have",而译文二使用的动词词组"be charged with(被……充满/填满)"让译文显得更加正式得体,其中用来修饰该词组的副词"heavily"意味"to a great degree",表示程度较深,对应原文"大量的"这一含义。

二 动词转译为副词

(1) After this, though the dust-dosed maize continued to suffer from phosphorus deficiency, the wheat and chickpea plants perked up and grew to more than double the size of their undusted lab-mates.

译文 在这之后,尽管施了灰尘的玉米持续遭受缺磷的折磨,小麦和鹰嘴豆植株却活跃起来,长得比未施灰尘的同类植株大了两倍多。

分析 “continue to + 动词”强调的是一种持续的状态而非动作本身，因此，在英译汉时，和它搭配的动词也是汉语中受到强调的动词“遭受”，“continue”也随之翻译成为汉语副词“持续(性地)”。

(2) A problem for manufacturers is that lighter cars tend to be noisy.

译文 制造商们遇到的一个问题是，重量较轻的汽车往往噪声大。

分析 “tend + to + 动词”表示动作的倾向性，“tend”本身不作为汉译时的核心动词。原句中的“noisy”在翻译时保留了其形容词性，“tend to”则翻译为了副词“往往”与之搭配。

第五节 名 词 化

科技文本旨在以事实为基础记述客观事物。因此，同样的内容，普通英语习惯用动词或动词性结构来表达，科技英语则更加偏好使用名词或名词性结构。名词性结构就是以短语形式来表述，是科技英语文本用于提高客观程度的重要途径之一。名词化(nominalization)结构简洁、传达确切，其广泛使用不仅能更加凸显科技文本的客观性与逻辑性，有时也能体现科技翻译美学——简约。

但是，科技领域中名词成灾(Noun Plague)现象十分明显，有时反而化简为繁，显得迂回作态。20 世纪 70 年代末，全球范围内开始掀起简明英语风潮，此后的写作风格有所改变。简明英语是以读者为中心的传递信息的方式，即在了解读者需求和知识水平的基础上，使用日常语言和熟悉的词语，有效传递信息的方式，使用简明英语语言，并不意味着削减要传达的信息。过多地使用名词化，如该用“suggest”，却用了“give a suggestion”，可能会导致译文句子复杂，结构冗长。

英语中名词化的基本结构常以下列几种形式出现。

一 词性变换的名词化结构

在英译汉时，源自动词的抽象名词需要将其相应的动词意义译出，但有时也可将名词化结构译入汉语中。

(1) 插入一条斜线便可以纠正这一错误。

译文一 You can rectify this fault if you insert a slash.

译文二 Rectification of this fault is achieved by insertion of a slash.

分析 译文一使用“you”将原句中的隐形主语人称显化，并且将原句中的“插入”“纠正”两个汉语动词也分别翻译成了英语动词“rectify”“insert”，这样使得整个译文呈现出一定的动态性和主观性。在信息类和祈使类文本中，译文一是符合要求的。但在科技文本中，译文二通过“名词+介词”搭配名词结构，将原句两个汉语动词名词化——“rectification”“insertion”，更加合适。

(2)在监督、管理、检查工作情况时，要保证合乎规格的工艺和/或材料处于产出状态。

译文一 When supervising, managing, and inspecting job, make sure that qualified workmanship and/or materials are produced.

译文二 On-the-job supervision, management and inspection ensures that proper workmanship and/or materials are being produced.

分析 译文二中的三个名词“supervision”“management”和“inspection”都派生于动词，科技英语中这类派生名词较为常见。译文一句子结构松散，语言较口语化。对比两个译文，在保证准确传达原文意义的前提下，后者的措辞更简洁严谨，结构也更紧密。

二 动名词形式的名词化结构

(1) Ensure personnel in performing, planning, approving shipboard maintenance on SUBSAFE, nuclear, FBW SCS and Level I systems participate in QA training.

译文一 确保参与执行、计划和批准潜艇安全、核、FBWSCS和一级系统的船上维修的人员参加质量保证培训。

译文二 确保执行、计划和批准潜艇安全、核、FBWSCS和一级系统的船上维修的人员参加质保培训。

分析 原文使用了三个动名词，即“performing”“planning”和“approving”。译文一将这三个词处理为名词，并在译文中增译了“参与”一词，显得累赘、不必要。译文二则直接将这三个词翻译为动词，没有一味追求名词性结构，使得译文简明流畅。

(2)在大范围内均匀地操纵电荷很棘手，尚无人能成功地实现其融合。

译文一 The wide and uniform manipulation of charge is tricky, and no one has done a good job to make sure its success.

译文二 Manipulating charge uniformly across a wide area is tricky, and no one has yet done it well enough to achieve successful fusion.

分析 译文一将“大范围”和“均匀地”处理为并列的句内成分，与原文所传达的意义不符，虽使用了“the+名词+of...”这一常见的名词化结构，但后半句呈现口语化的表达特征。译文二则是直接使用动名词“manipulating”，句子结构一目了然，前后文本正式程度保持一致。

(3)如果提高电流的频率，就会使这种运动速度加快。

译文一 If the frequency of the current is raised, then the movement will be sped up.

译文二 Raising the frequency of the current speeds up the movement.

分析 对比两个译文,前者照应原文翻译为由"if"引导的条件状语从句,两个小句均以被动语态形式出现;但后者直接使用动名词"raising",抓住了主要动作,这让译文结构简单清晰,表达通畅,更有利于读者获取关键信息。

三 其他形式的名词化结构

(1)他们会发现,这些系统不仅安装成本和使用成本高,而且毫无必要地消耗着非再生能源,排放出温室气体。

译文一 They will find that installing and using these systems are expensive, and they necessarily consume non-renewable energy and emit greenhouse gases.

译文二 They will find these systems expensive to install, costly to run, and are unnecessary consumers of non-renewable energy and emitters of greenhouse gases.

分析 对比两个译文,译文一对"使用"一词的翻译不准确,同时前后两个小句都将"they"用作主语,句子层次不清晰。译文二中"expensive to install"和"costly to run"结构清晰,指向明确,便于译文读者理解。在此译文中,"consumers"和"emitters"分别对应动词"consume"和"emit",这是因为一些有显著动作意义的名词与"of"短语配合使用时,表面上指某个施事者,但其中的动作意义往往需要在译文中来体现。这样形成前后一致的对称结构,整个译文信息传达确切,结构严密。

(2)人们曾相信科学和技术的方法是获得真理和知识的唯一有效途径。然而,这种信念正在改变。

译文一 They once believed that scientific and technological methodologies are the only valid way to gain truth and knowledge. However, this belief is changing.

译文二 Their confidence that scientific and technological methodologies are the only valid approaches to truth and knowledge is changing.

分析 对比两个译文,在处理"是获得……的途径"这一结构时,译文二将其翻译为"approach to...",使译文简洁明了;同时,使用"Their confidence that...is changing."同位语从句,避免了如同译文一那样的重复翻译,句子较严密,意义传达也清晰。

(3)如果控烟干预不能够实施的话,由吸烟导致的死亡人数会持续增长。

译文一 If the restriction on smoking fails to implement, the death toll will continue to increase.

译文二 Failure to implement restrictions on smoking will result in an increase in smoking-related death rates.

分析 原句是一个汉语条件句,译文二在英译时采用名词化结构处置,将"如果"引导的条件从句用作主语，译成名词化结构“failure to...”，表示事情的起因；主句作宾语，译成名词化结构“an increase in...”，表示事情的结果；时态为将来时态。译文一遵照原文的叙事逻辑，翻译出了条件状语从句，在结构和逻辑上反倒不如译文二顺畅。

科技文本需要保证客观性与简洁性，因此多使用名词或名词性结构。但在其他领域进行英语写作时，提倡尽可能使用动词，而不是名词化的形式。因为许多名词化的句子通常以“be”动词作为主要动词。使用以名词化形式的动词作为主要动词，有助于行文表达引人入胜而非令人沉闷。

练习题

一、英译汉

1. Even a protective facility is no insurance against death from lack of oxygen.

2. The circuits are connected in parallel in the interest of a small resistance.

3. Black holes act like huge drains in the universe.

4. All metals tend to be ductile.

5. These parts must be proportionally correct.

6. Gene mutation is of great importance in breeding new varieties.

7. Earthquakes are closely related to faulting.

8. Each sample must be submitted with full particulars of its source.

9. The computer is a far more careful and industrious inspector than any human being.

10. However, incomplete community knowledge of the drug's proper use suggests that education efforts may further improve outcomes.

二、汉译英

1. 操控国外生产的汽车是需要懂一些英语的。

2. 我们确信这个实验会成功。

3. 人们能利用智能手机联系千里以外的朋友。

4. 这些材料的特点是结构紧凑、携带方便。

5. 现代天气预报的准确度很高。

6. 任何人不能违背各国人民的利益去人为地挑起紧张态势，甚至制造武力冲突。

7. 数以百万计的非洲人已逐渐意识到他们的生活状况异常贫穷落后，这就促使他们奋起采取坚决措施去创建新的生活条件。

8. 为有效控制汽车排放，减少汽车尾气对大气环境的污染，就必须精确控制混合气空燃比。

9. 全球人口快速增长，耕地资源日渐稀缺，农产品市场迎来了全面的牛市。

10. 加热面上的迅速蒸发，往往使蒸汽的湿度加大。

第八章
反译

由于不同的语言有不同的表达习惯和语言特点,即使是描述同一种情景或同一个意思,也可能会出现相反的表达方式(英语和汉语两种语言也不例外)。为使译文更真实地反映原文语言环境的表达习惯,文学翻译中的反译法也常用于科技文体的翻译实践中。所谓反译,就是指在翻译时采用与原句相反的语序或表达方式来呈现原句的意义。即不拘泥于原句的表达形式,将从正面表达的句子,在目的语中通过反向表达的方式来传递原句内涵,反之亦然。反译法的使用可使译文更符合目的语的表达习惯和用语规范。

在科技翻译中,反译法是双语转换时常见的翻译方法,通常为否定形式和肯定形式之间的同义替换。受历史、文化、社会等因素的综合影响,不同的语言在构建语言表达式时,在词语、短语选择和理解等方面各有其章法和习惯,因此在保留原文信息的基础上进行合理反译,不仅有助于使译文通顺明晰,信息透彻传达,还有助于降低目的语读者的理解负担。译者首先应正确理解原文含义,厘清原文逻辑和写作目的。然后判断是否需要通过反译法,将原文进行肯定与否定相互转换。最后,使用符合目的语表达习惯的表达式传递原文信息,切忌生硬翻译。反译法有时也会延伸到主动与被动、静态与动态(包括词性转换等)、句子顺序等方面的翻译中,因为本教材在其他章节对以上内容都进行了单独的讲解,所以不在本章重复介绍。本章将从肯定与否定相互转换的角度展开分析,主要分为显性否定、隐性否定和显性肯定三种情况。

第一节　显性否定

在汉语和英语中,表示否定意义的表达式有很多,其中大部分为显性否定。显性否定通常指在语言形式上有着明显否定标记的表达。一些常见的英文否定词有"no, not, none, never, neither, hardly"等,或带有否定意义的前缀或后缀,如"dis-, non-, im-, mis-, un-, anti-, -less"等。汉语中常见的否定词有"没、弗、不、无、勿、休、别、莫、非、并非"等。在双语转换中,常会出现通过否定形式表达肯定意义的情况,但如果采取顺译方法,则会面临源语中的一些否定表达可能无法与目的语自身表达习惯相契合而造成误译或生硬翻译的情况,因此,应正确剖析原文的内涵,再通过反译法进行转换。

(1) Gasoline is not as dense as water, so it floats on water.

译文一 汽油的密度不如水,因此可以漂浮在水面上。

译文二 汽油的密度比水小,因此浮在水面上。

分析 “not”为最常见的构成显性否定结构的英文否定标记词之一。译文一保留了否定形式“不如”,但是表达模糊,未能正确传达原文的信息。译文二用肯定形式表达了否定的意义,用“比”直接引出比较的对象,突显了汽油密度更小的信息。另外,密度是对特定体积内的质量的度量,相较于原文,译文二用汉语中常用的密度衡量表达方式准确传递了其“大”或“小”的信息。

(2) Anti-aircraft gun shells are usually fitted with special fuses that break apart in the air if they reach a certain height or time.

译文 高射炮的炮弹通常装有特殊的引信,只要达到一定高度或时间,就会在空中爆炸。

分析 若将“Anti-aircraft gun”直译为“反飞机大炮”,则明显不符合汉语中的常见表达习惯。忽略否定形式,剖析其深层含义,我们可以得知“Anti-aircraft gun”主要指的是用于攻击各类空中目标的军事武器。在汉语中,通常从正面将其表达为“高射炮”或“防空炮”。

(3) 因为塑料不导电,所以常用作绝缘材料。

译文一 Plastic is often used as insulation materials because it does not conduct electricity.

译文二 As plastic prevents electricity passing through it, it is often used as an insulating material.

分析 两则译文均传达出原文中“不导电”的否定含义。译文一的表述更为绝对,用显性否定词直接否定了塑料的导电能力。译文二用肯定的形式突显了塑料具有一定的阻止电流通过的能力。结合科学事实,早在2000年,诺贝尔化学奖就被授予了导电塑料的研究者,因此当前塑料是否完全不具有导电能力,还无法一概而论。译文二用肯定形式表述了具有否定含义的信息,“prevents”相比译文一中的“does not conduct”更留有一定余地,而非完全否定,更具尊重科学事实的意味。

(4) 但是该团队还没有找到可行的方法来做到这一点,而且他们还需要想办法将这项技术拓展到实际应用中。

译文一 But the team hasn't found viable ways to do that yet. They will also have to figure out a way to scale the technology for real-world applications as well.

译文二 But the team has yet to come up with viable ways of doing so. They will also have to figure out a way to scale the technology for real-world applications as well.

分析 译文一使用了显性否定词“not”,强调至今仍然“没有”找到可以实现该项技术的途径,其否定意味十分强烈。译文二使用了暗含否定意义但具有肯定形式的词

"have yet to do",意为"还(尚)未",其中之"暂未"意思不仅传递了技术研发尚未成功的深层信息,而且暗含着对该项技术未来发展的期待。

第二节 隐性否定

隐性否定是一种形义错位的特殊否定形式,通常指在语言形式上没有否定标记,但其词汇意义或引申意义具有否定内涵的表达。英语中的隐性否定有"avoid, lack, refuse, out of, short of, ignorance, exclusion, too...to..., rather than"等表达,在汉语中也有"差点、抵制、防止、难免"等。在翻译时,我们可以将隐性否定看作通过肯定形式表达否定意义,需要根据原文内容将其由肯定形式转换为否定形式。

(1) Although bacterial bioluminescence genes can be targeted to plastids to engineer autoluminescence, it is technically cumbersome and fails to produce sufficient light. The caffeic acid cycle, which is a metabolic pathway responsible for luminescence in fungi, was recently characterized. We report light emission in Nicotiana tabacum and Nicotiana Benthamian plants without the addition of any exogenous substrate by engineering fungal bioluminescence genes into the plant nuclear genome.

译文一 虽然生物发光细菌基因可以针对质体设计自发光,但技术很复杂,且失败于产生足够的光。真菌发光的代谢途径最近被确定为咖啡酸循环,我们通过将真菌生物发光基因导入植物核基因组就使烟草和本氏烟发光了,无须添加任何外源底物。

译文二 虽然生物发光细菌基因可以针对质体设计自发光,但技术很复杂,发的光也不够。真菌发光的代谢途径最近被确定为咖啡酸循环,我们通过将真菌生物发光基因导入植物核基因组就使烟草和本氏烟发光了,无须添加任何外源底物。

分析 从形式上来看,"fails to"不含否定标记,但在此处具有否定意义,表示"未能,未做成"。如果将其直译为"失败于产生足够的光"则显得十分生硬,因此使用反译法将其翻译为"不够"更符合中文的表达习惯。同样,"without"不是显性否定词,但带有否定意义,在此处是"没有,不借助"的意思,翻译为否定形式"无须"使得译文更加流畅且易懂。

但当"without"在"The researchers synthesized genes located in two different classes of protein-polymers without delay."(研究人员立即将位于两个不同层级的蛋白质聚合体的基因进行合成。)一句中时,则无须译为具有否定标记的显性否定表述。"without delay"原指"没有任何延迟",在此处将其译为"立即",相比译为"没有迟疑"更具有紧迫性,在贴合原文上下文语境的同时,更符合科技文体用语简洁凝练的特点。

(2) Avoid letting the temperature of your battery-powered gadgets dip below 32 degrees Fahrenheit or soar above 95 degrees.

译文一 避免将以电池供电的设备置于32华氏度(0摄氏度)以下或95华氏度(35摄氏度)以上的温度环境中。

译文二 不可将以电池供电的设备置于32华氏度(0摄氏度)以下或95华氏度(35摄氏度)以上的温度环境中。

分析 译文一遵循了原文中"avoid"的正面表达形式,译文二则是从反面入手,使用了否定词"不可"。"不可"二字使得译文二更具有警示意味,提醒读者"禁止"采取不当操作,否则可能带来危险。而译文一更像是一种温馨提示,带有可遵循亦可不遵循的暗示,读者难以直接获取原文想要传达的警告意思。

(3) Repeating the trick in humans is far from straightforward, but the findings will fuel interest in radical new therapies that aim to slow or reverse the ageing process as a means of tackling age-related diseases such as cancer, brittle bones and Alzheimer's disease.

译文一 尽管用同样的方法让人类返老还童仍任重道远,但是这一发现会点燃人们对以延缓或逆转衰老为目标的激进新疗法的兴趣,这种疗法可以用来治疗癌症、骨质疏松和老年痴呆症等老年常见疾病。

译文二 尽管用同样的方法不能让人类返老还童,但是这一发现会点燃人们对以延缓或逆转衰老为目标的激进新疗法的兴趣,这种疗法可以用来治疗癌症、骨质疏松和老年痴呆症等老年常见疾病。

分析 译文一遵从了原文的表达形式,从正面说明在当前的科学技术下,该疗法无助于人类返老还童。相比之下,译文二由肯定转换为否定的表述更加直接明了,否定词"不能"强化了否定意味,直指当前疗法的局限性。且译文二更加符合科技文体用词简洁准确的特征。

(4) The number of enigmatic bursts detected so far is too small to draw any conclusions.

译文一 目前探测到的快速射电暴太少,未得出任何结论。

译文二 目前探测到的快速射电暴太少,无法得出任何结论。

分析 "too...to"结构是英语中的固定搭配,是典型的以肯定形式暗含否定意义的表达。如果直译则十分生硬,不符合汉语惯用表达,降低了译文的可读性,因此在汉语中通常反译为"太……而不能"。

(5) 这种双模车顶多能搭载21名乘客。

译文一 The dual-mode vehicle can carry at most 21 passengers.

译文二 The dual-mode vehicle can carry no more than 21 passengers.

分析 "顶多"指"最多不过",强调的是实际数不会高于该最高数值,该词为肯定形式但含有否定意义。相比直译为译文一中的"at most",译文二中的"no more than"从否定的角度传达原文含义,强调承载量不可超过21人的信息。

(6)因火星大气中氧气含量约为0.14%,所以人类一直都认为火星上的稀薄氧气难以维持微生物的生命。但是本次的新发现彻底改变了我们对火星上可能存在生命的认识。

译文一 The oxygen content of Mars atmosphere is about 0.14%, so it has long been assumed that the trace amounts of oxygen on the Red Planet are hard to sustain any microbial life. But the new discovery fully revolutionizes our understanding of the potential for life on Mars.

译文二 The oxygen content of Mars atmosphere is about 0.14%, so it has long been assumed that the trace amounts of oxygen on the Red Planet cannot sustain any microbial life. But the new discovery fully revolutionizes our understanding of the potential for life on Mars.

分析 "难以"一词不具有否定形式,但隐含否定意义。译文将其译为否定形式"cannot",强化了否定意义。结合后文内容,可以判断出新发现与以往的人类经验是背道而驰的,因此,在此处采用完全否定可与后文形成更鲜明的对比,以突显新发现的重要性,使译文更加通顺合理。

第三节 显性肯定

除以上两种情况外,有时原文为显性肯定形式,即在表层结构上没有否定词,且在深层结构中也没有否定意义的暗示。因此,一方面为使译文更符合目的语表达习惯,另一方面为突显原文作者写作意图,译者需在保留原文信息内涵的基础上,合理地将肯定表达反译为否定表达。

(1) According to the findings, the flames left the red wood trees charred but still mostly alive. The devastating fire swept through California's Big Basin Redwood State Park and the rest of the life that usually animates the forest was gone.

译文一 调查结果显示,大火将红杉树尽数烧焦,但它们大部分都还活着。这场毁灭性的大火席卷了加州大盆地红杉州立公园,那些平常让红杉林生机勃勃的其他生命也都消失了。

译文二 调查结果显示,大火将红杉树尽数烧焦,但大多数没烧死。这场毁灭性的大火席卷了加州大盆地红杉州立公园,那些平常让红杉林生机勃勃的其他生命却消失殆尽。

分析 译文一保留了“alive”的肯定形式和肯定内涵“活着”，从正面传达了红杉树仍然大部分存活的信息，忠实于原文的表达形式和内容。译文二通过反面的表述“没烧死”来描述这一场景，与上半句中的“烧焦”形成对比，更突显了火灾的悲剧性和红杉树生命力的顽强。

(2) Researchers are cautious about using Yamanaka factors in humans because previous work has shown that fully reprogrammed cells can turn into clumps of cancerous tissue called teratomas.

译文一 研究人员对在人类身上使用山中因子持谨慎态度，因为先前的研究表明，经过完全基因重组的细胞会转变为名叫畸胎瘤的癌变肿块。

译文二 科学家不轻易给人类注射山中因子，因为先前的研究表明，经过完全基因重组的细胞会转变为名叫畸胎瘤的癌变肿块。

分析 “be cautious about”意为“对……小心”。译文一结合上下文语境，保留了原文从正面表述的形式，将其译为“持谨慎态度”。译文二则直接从反面表述，使用了具有否定意义和形式的“不会”。相比于译文一，译文二杜绝了模棱两可的态度，使信息的传递更加直接和简洁。

(3) While the social distancing innovation is new, the research company has deployed more than 10,000 sensors for workspace optimization, including in the offices of Facebook and Dell.

译文一 尽管这一社交隔离创新项目还是新的，但是该研发公司已在脸书、戴尔等大型企业的办公室安装了超 1 万台传感器来优化办公空间利用率。

译文二 尽管这一社交距离创新项目推出不久，但是该研发公司已在脸书、戴尔等大型企业的办公室安装了超 1 万台传感器来优化办公空间利用率。

分析 原文中的“new”为肯定形式且不含有否定意义，在英语语言环境中这样的表述十分常见。但在汉语中，如果将其翻译为译文一中的“新的”，则明显不符合科技文体去口语化、信息明晰简洁的特点。另外，即使在汉语中这样的表述也并不通顺，略显生硬。译文二从反面入手，将其译为“推出不久”，则更加符合科技文体的特点，也更加符合目的语环境下的表达习惯。

(4) The company, Genomic Prediction, already offers a test aimed at screening out embryos with abnormally low IQ to couples being treated at fertility clinics in the US.

译文一 这家名为 Genomic Prediction 的公司已经为在美国生育诊所接受治疗的夫妇提供了一项检测服务，旨在筛查出智商异常低的胚胎。

译文二 这家名为 Genomic Prediction 的公司已经为在美国不孕不育诊所接受治疗的夫妇提供了一项检测服务，旨在筛查出智商异常低的胚胎。

分析 “fertility clinics”在柯林斯词典中的释义为“a place at which a couple who are unable to conceive may receive medical advice and treatments to help them to have a child”(一个可以为不能怀孕的夫妇提供医疗建议和治疗,帮助他们生育的地方),主要提供如试管婴儿助孕等服务,字面意思为“生育诊所”。但在中文中,鲜少有“生育诊所”的说法,对这类专治“不孕不育症”(超过一年无法成功妊娠的不孕症和不育症的统称)的医院或诊所,常用“不孕不育医院”之类的名词称呼。因此,将“fertility clinics”由肯定形式转换为否定形式的“不孕不育诊所”更加符合中文的表达习惯,补全了原文暗含的信息,减少读者认知负担,可以更为直接地获取原文含义。

(5)自然界中的作用力,必定要牵涉到二个或二个以上的物体。

译文一 Every force acting in nature must involve two or more bodies.

译文二 There is never a force acting in nature unless two or more bodies are involved.

分析 原文是一个陈述句,不带有否定词或深层的否定含义。译文一采用了直译的方法,保留了原文的肯定形式。译文二采用了反译法,增加了表示否定的副词“never”与其后的“unless”搭配,更具强调作用,使句意更加明晰。相比译文一中的“every force”,“never a force”语气更加强烈,更加突显自然界中的所有作用力都具有以上特点。

练习题

一、英译汉

1. Few people would support research on animal hybridization.

2. Experimental data and expected data are badly out of line.

3. Modern vessels, fitted with radar installations, are safe from collision with other vessels.

4. This is the point at which a quantum computer is shown to be capable of performing a task that is beyond the reach of even the most powerful conventional supercomputer.

5. It is often very energy consuming to produce the enzymes that inactivate the antibiotics.

6. The research project was forced to stop because the gas explosion experiment had gone beyond control.

7. The Pfizer treatment could help keep people infected with the coronavirus from getting so sick that they need to be hospitalized.

8. The map showed that the galaxy's disk, far from flat, is significantly warped and varies in thickness from place to place, with increasing thickness measured further from the galactic center.

9. And beyond being able to more precisely understand how many mountain lions are in an area, Alexander says that this new camera trap method could be used for tracking other critters that lack distinguishing side colors but have unique features elsewhere.

10. On the other hand, commercial apps usually aren't designed for research, which demands predictable, transparently collected, and granular data. Sometimes, that means app-generated information is actually less useful to researchers.

二、汉译英

1. 所有的电磁波在真空中都具有相同的波速。

2. 电气栽培对蔬菜和站在一旁的工作人员完全无害。

3. 随之而来的化学反应非常复杂,不出所料的话,混凝土会变得越来越硬。

4. 该种植室使用人造光源和电场来刺激植物生长和预防疾病,操作自动化,几乎不需要护理和维护。

5. 科学家认为月球上的水是以水分子的形式存在的,它们与月球尘埃结合在一起,而不是以表面的冰层形式存在。

6. DeepMind 的科学家开发了一个系统,它仅依赖自身图像传感器所输入的信息就可以学习,且无须人类监督。

7. 压电式对管道振动十分敏感,但在一定范围内,差压式几乎不受管道振动干扰的影响。

8. 目前还远不能证明冻结神经能永久性减轻体重,但如果确实如此,这可能会对那些努力保持健康体重的人的生活产生深远影响。

9. 另外两名接受植入手术的截瘫患者也能在不同程度上活动他们的腿,其预后前景非常乐观。

10. 同十年前的情况大不相同的是,基于高通量测序的液体活检技术的应用大大提高了癌症检测和早期干预的可能性。

第三部分

科技翻译中的常见表达

第九章 数字的翻译

数字是一种计数符号,也是科技文本中使用频次较高的基础信息载体。科技文本中数字所包含的信息量非常大,因此,精准客观地翻译数字,有助于更加精准地阐释和说明医学、化学化工、信息技术等领域的科学信息。但中西方的思维习惯和表达方式存在诸多差异,因此英语和汉语关于数字的计数习惯和描述方式并不完全一致,在写作上存在错位或空缺的情况。如在汉语中通常四位数为一级,而在英语中通常三位数为一级。这就导致在翻译四位数以下的数值时可以采用直译,翻译四位数以上的数值则需遵循目的语的数字表达规则。另外,英语中超过三位数的数字,通常以小数点为中心,整数部分从右到左每三个数加一个千分撇(如 66,666.6),汉译中可以(不是必须)使用千分撇(如 66,666.6)或千分空(66 666.0)。再者,英译汉过程中有时会面临译为阿拉伯数字还是中文数字的抉择,遇到这种情况,我们一般参照《出版物上数字用法》(GB/T 15835—2011)中的相关规范。阿拉伯数字具有醒目和易于辨识的特点,因此多用于计量(长度、面积、体积等)、编号(电话号码、道路编号、产品型号等),并且有部分含有阿拉伯数字的词语是已定型的,如 92 号汽油、5G 手机、“5·12”汶川大地震等。而在面对以下三种情况时,多选用中文数字的写作形式:非公历纪年(干支纪年、农历月日、历史朝代纪年等)、概数(三四个月、几万分之一、四五万套)和已定型的含汉字数字的词语(三叶虫、一氧化碳、七七事变等)。

本章将从基数和不定数、分数和小数、数的增减、倍数的增减、数学运算的翻译五个小节介绍科技文本中常见的数词类型和数值翻译方法。

第一节 基数和不定数

一 基数

在数学中,基数是用来描述一个集合大小的概念。广义上来说,基数就是用来表示数目、具有计数功能的自然数。中英文的读数法存在较大区别,中文数字分区一般为四个区间,第一区间包含“个、十、百、千”,第二区间包含“万、十万、百万、千万”,第三区间包含“亿、十亿、百亿、千亿”,第四区间包含“兆”(即万亿)。英文数字分为五个区间,第一区间为“one,ten,hundred”,第二区间为“thousand,ten thousand,hundred thousand”,第三区间为“million,ten million,hundred million”,第四区间为“billion,ten billion,hundred billion”,第五区间为“trillion”。即便大多数时候中英数字都可以用阿拉伯数字表示,但也存在数值或单位换算的情况。在翻译基数词的时候,一般采用直译法和转换法两种方法。

(一)直译法

当数值相等且在目的语和源语中写法相对应时,可以直接保留原文中的数值,同时直译原文中的数字单位。

(1)The total bandwidth of the ground network is about 20 kilobits per second.

译文 地面网络的总带宽约为每秒 20 千比特。

分析 千比特/秒也常写作 kbps,指数字信号的传输速率。该例中数值较小,不涉及进制转换和英汉单位对等,因此可以直译。

(2)A single battery offers a talk time of up to 10 hours.

译文 单块电池的最长通话时间可达 10 小时。

分析 英汉两种语言中的时间单位一致,分为时(hour)、分(minute)、秒(second)。因此,这里采用直译法,保留原文中的数值写作方法和时间单位。

(3)By comparison, researchers hope to be able to bring the cost of a solar road to $ 310 to $ 460 a square meter with mass production.

译文 相比之下,研究者期望在大规模生产时能把太阳能公路的成本降低到每平方米 310 美元到 460 美元。

(4)A pulse of neutrinos (small, elusive subatomic particles with no electric charge) corresponds to the digit “1” while no pulse to “0”.

译文 一个中微子(一种质量轻、不带电荷且难以捕捉的亚原子粒子)出现脉冲时对应数字“1”,没有脉冲时则对应数字“0”。

分析 例(3)和例(4)都可以直译,不涉及单位或数值的换算。且在以上英译汉例句中保留阿拉伯数字的形式,而非处理为“一”或“零/〇”,即可达到易于辨识数值的效果。

(二)转换法

转换法通常用于双语中数值不对等或单位不一致的情况,译者可通过转换法对数值或单位加以换算,以符合目的语中的数字表达习惯。

(1)The project leader put the total price tag for the finished segment at about $ 10 million, including the cost of excavation, internal infrastructure, lighting, ventilation, safety systems, communications and a track.

译文 项目负责人将这段完工隧道的总造价定在1000万美元左右，其中包括挖掘、内部基础设施、照明、通风、安全系统、通信和轨道的成本。

分析 按照英语习惯，一千万用阿拉伯数字写作10,000,000，意为十个一百万，即10 million。但中文习惯与之不同，中文采用的是万位进制，一千万用阿拉伯数字写作1000,0000，即1000万。

(2) The research shows that a drop of 100 microlitres of water released from a height of 5.9 inches can generate a voltage of over 140V, and the power generated can light up 100 small LED lights.

译文 研究表明从约15厘米高处滴下的100微升的水发电量超过140伏，能点亮100个小的LED灯。

分析 "inch"(英寸)是英语中常用的长度单位，1英寸约等于2.54厘米。按照中文计数习惯，经过数值和单位的转换，译为"约15厘米"。

(3) The Earth is struck by lightning nearly 20 million times each year, and bolts of lightning can travel as much as 10 to 12 miles from a thunderstorm, instantly heating the air to 50,000 degrees Fahrenheit, according to the National Weather Service.

译文 根据美国国家气象局的数据，地球上每年发生的闪电次数可达近2000万次，雷雨过程中闪电可行进10~12英里(16~19千米)，瞬间让空气升温5万华氏度(27760摄氏度)。

(4) The rover, built in NASA's Jet Propulsion Laboratory in Pasadena, Calif., is about 10 feet long, 9 feet wide, 7 feet tall and about 2,260 pounds.

译文 这个火星车是在美国宇航局位于加州帕萨迪纳市的喷气推进实验室制造的，车体长约10英尺(约3米)，宽约9英尺(约2.7米)，高约7英尺(约2.1米)，重约2260磅(约1025千克)。

(5) In the confined space of an Easy-Bake oven, a 100-watt bulb can create a temperature of 325 degrees Fahrenheit.

译文 在一个封闭空间——比如Easy-Bake烤炉里，一只100瓦的白炽灯泡可以达到325华氏度(约163摄氏度)的高温。

分析 译者还可以在译文中保留原文数字，补充转换后的数字。如例(3)~例(5)中，英里和千米、华氏度和摄氏度、英尺和米、磅和千克都是对同一客观数值的不同计数单位，因此为减轻目的语读者的理解负担，通过转换法对其进行换算很有必要。

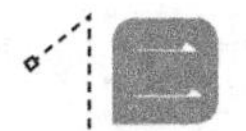

二 不定数

不定数也可以称为概数，表示大概的、模糊的数值。根据具体情况，还可以将它们分为约等于、小于和大于某一数值三种情况。

(一)约等于某一数值

在英文中,常在整数前使用"about,around,roughly,approximately,some,close to,more or less,in the neighborhood of"等词或短语,或将"or so,or thereabout,in the rough"等短语置于整数之后,表示大约、非确切的数值。在中文中,则常使用"大约、左右、近似、上下、近乎、接近"等词来描述这类数值。

(1) The panchromatic images were taken by the high-resolution camera of Tianwen-1 at a distance of 330 to 350 km above the surface of Mars, with a resolution of about 0.7 meters. It is estimated that the diameter of the largest impact crater in the images is around 620 meters.

译文 全色图像由天问一号高分辨率相机在距离火星表面330～350千米高度拍摄,分辨率约0.7米。据测算,图中最大撞击坑的直径约620米。

(2) Prior to the orbital correction, the Chang'e-5 lunar probe had traveled for roughly 17 hours in space, and was approximately 160,000 km away from Earth. All of the probe's systems were in good condition.

译文 截至第一次轨道修正前,嫦娥五号探测器各系统状态良好,已在轨飞行约17个小时,距离地球约16万千米。

(二)小于某一数值

在英文中,常在整数前使用"under,below,within,fewer than,less than"等词或短语,表示比某一数值小。在中文中则通常使用"小于、少于、低于、不超过"等词。

(1) The prototype is around 5 feet tall thanks to the roll bar and a little less than 12 feet long and 12 feet wide.

译文 由于有翻车保护杆,原型机高约5英尺(约1.5米),长不足12英尺,宽12英尺。

(2) Vibrations in the spacecraft below 10 Hz can damage the internal organs, and even threaten a person's life。

译文 10赫兹以下的低频振动会引起人体内脏共振,甚至会危及其生命。

(三)大于某一数值

在英文中,常在整数前使用如"more than,over,above,upwards of",或在其后接"and more,odd,and odd"等词语或短语,表示比某一数值大。在中文中,则通常使用"大于、超过、远超、多于、高于"等动词搭配数字及其单位,置于分句末。还可以用"余""多"(置于数量词后)表示零头。

(1) Lynx spacecraft can travel at a speed of more than 2,500 mph—and dozens of miles above

the earth—before safely landing at an airport.

译文 “山猫”号航天飞机时速可超过2500英里,在机场安全降落前飞行高度也可达到几十英里。

(2)腾讯QQ即时信息服务拥有7亿多名用户,该服务有许多附加功能,比如有可以让用户改换形象的“皮肤”。

译文 Tencent's QQ instant messaging service, which has more than 700 million users, offers additional features such as skins to change the look of the client.

分析 若将例(2)回译到汉语,不少译者会处理成“拥有超过某数字的名词”的句型,比如“腾讯拥有超过7亿名的用户”,这个中文表达受到“more than”的负迁移,不符合汉语语法习惯。正确的表达为“用户超过/逾7亿名”或“拥有7亿多/余名用户”。

(四)固定表达

所谓固定表达,是指那些不精准确切的数值需要通过已约定俗成的固定表达来描述。如英文中的“ten/ dozen/ score/ hundred/ thousand 的复数形式 + of”,中文中的“数十”“数百”“数千”等词或短语。

数十/几十:tens of; dozens of; scores of; decades of

数百/数以百计/几百/成百:hundreds of; several hundreds of; several hundred; by the hundred

数万/数以万计/好几万:tens of thousands of

数百万:millions of

数千万:tens of millions of

数亿:hundreds of millions of

数十亿:billions of

数百亿:tens of billions of

数千亿:hundreds of billions of

数万亿:trillions of

第二节 分数和小数

在数学中,单位“1”可以等分为若干等份,表示其中的一份或几份的数则为分数。分数由分子、分母、分数线组成。英文分数的分子和分母分别由基数词和序数词表示,分子大于1时,序数词分母加s,如 seven tenths(7/10)。百分数则表示为“数字 + percent”,如25%可以表示为25 percent。中文分数通常表示为“几分之几”,如六分之一、十二分之五。

小数由分数转化而来,也是实数的表现形式之一。小数由整数、小数和小数点组成,以小数点为界,左边为整数部分,右边为小数部分。翻译小数时,通常采用直译法。

(1) Based on early experiments and calculations, the plate approach promises a 639 percent increase in strength and a 522 percent increase in rigidity over the beam nanolattice approach.

译文 基于早期的实验和计算,这种平板结构与束纳米晶格相比,强度会提高 639%,硬度会提高 522%。

(2) Once folded, the CityCar will fit into a space just one-third the size of a standard parking spot.

译文 折叠后的 CityCar 只占标准停车位三分之一的空间。

(3) According to the Environmental Working Group, 75 percent of 800 sunscreens tested in the US contained potentially harmful ingredients. Only one-fourth of them were effective at protecting our skin without any toxicity.

译文 据环境工作小组报告,美国送检的 800 种防晒产品中,有75% 含有潜在危害成分。仅有1/4 的产品既不含毒素,又能有效保护皮肤。

(4) 轮胎的滚动阻力系数与花纹和材料相关,一般介于 0.015 到 0.02 之间。

译文 The rolling resistance coefficient of a tire depends on the pattern and material, and is generally between 0.015 and 0.02.

第三节　数的增减

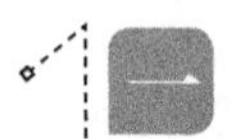

一　数的增加

表示数的增加的常用词和固定搭配有很多。例如,"增"为"rise, go up, grow, increase/a rise, an increase"等;"猛增"为"hike, jump, shoot up, soar, skyrocket, surge/a hike, a jump, a surge"等;"缓增"为"climb, pick up/a climb, be on the increase"等;"创新高"为"hit a record high, scale new heights, reach a peak"等。

(1) By burning coal, oil and gas, humans have increased carbon dioxide amount to 385 ppm; it continues to grow by about 2 ppm per year.

译文 人类燃烧煤炭和油气的行为使二氧化碳含量上升至 385ppm(百万分之三百八十五),而这一数值还在以每年约 2ppm(百万分之二)的速率增长。

(2) Although the temperature of the furnace has gradually risen to 200℃, it is still unable to

fully burn the material.

译文 尽管实验炉温度已逐渐升至200摄氏度,仍然不能将该材料化为灰烬。

(3) The market share of Samsung's smartphones has almost tripled to 34%.

译文 三星智能手机的市场份额几乎增加了两倍,达到34%。

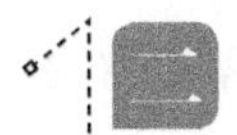

数的减少

表示数的减少的词和常用搭配也有很多。其中"降"为"fall, go down, drop, reduce, decline, decrease/a fall, a drop, a reduction, a decline, a decrease"等;"猛降"为"plummet, plunge, slash, tumble/a plunge"等;"稍降"为"dip, slip, trim/a dip";缓降,"be on the decrease/decrease"等;"创新低"和"跌入谷底"常用"hit a record low, hit the bottom"等表述。

(1) We have shown that muscle activity in the back, shoulders and knees drops by 50%. If muscle activity drops, that means the risk of muscle injury is less.

译文 我们已证明,背部、肩膀和膝盖的肌肉活动减少了50%。肌肉活动减少,意味着肌肉损伤风险降低。

(2) Light moves along at full "light speed"—186,282.4 miles per second—only in a vacuum. In the dense matrix of a diamond, it slows to just 77,500 miles per second.

译文 光只在真空中才以"光速",即186,282.4英里/秒(299,792,458米/秒)运动。在钻石这样的高密度物质中,光速会下降到77,500英里/秒(124,724,160米/秒)。

其他表示数量在某一范围波动或持平的固定表达包括达到平衡(to level out/off)、保持平衡(to remain stable)、保持在(to stand/remain at)、达到高峰(to reach a peak)等。

第四节 倍数的增减

倍数翻译一直是英汉翻译中的重点和难点,因为受语言和思维差异的影响,英语和汉语中倍数的表达结构各异,类型多样。

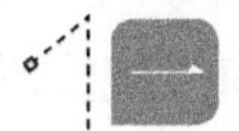

倍数增加

除了一些常见的表述倍数增加的词,如"double"(翻番)、"quadruple"(翻两番)和"triple"(增至三倍/增加两倍)外,常用的描述倍数增加的句型包括:

(a) A is *N* times as great (long, much, ...) as B;

(b) A is *N* times greater (longer, more, ...) than B;

(c) A is N times the size (length, amount, ...) of B.

以上三种句型均应译为"A 的大小(长度,数量,……)是 B 的 N 倍",或者"A 比 B 大(长,多,……)$N-1$ 倍"。其中,句型(b)是难点也是易错点。

在 Jim Loy 的化学元素周期表中有这样一个句子:An oxygen atom is almost exactly 16 times heavier than a hydrogen atom. It would seem that oxygen is made up of 16 nearly equal pieces (particles), while hydrogen is just made up of one。学过化学的都知道,氧原子质量为 15.9994(约 16)u,氢原子质量为 1.00794(约 1)u。这就说明在上述(b)句型中的"greater/longer/more"仅表示 A 比 B 大/长/多,但与中文不同的是,此处的数字是多少就表示是多少倍。

(1) The oxygen consumption of experiment A is three times as big as (three times bigger than, three times the size of) experiment B.

译文 实验 A 的耗氧量是实验 B 的 3 倍(即大两倍)。

(2) While today's best digital cameras take images having pixel counts in the tens of millions, the latest device produces a still or video image with a billion pixels, which is five times more detailed than can be seen by a person with 20/20 vision.

译文 现今最好的数码相机可拍摄分辨率达数千万像素的图片,这款最新的相机能够拍摄 10 亿级像素的静态图片或视频图像,而且清晰度比正常裸眼视力看到的图像要清晰 4 倍。

分析 "five times more"的意思为"是/有 5 倍那么(形容词)",这里即"有……看到的 5 倍那么清晰",也就是"比……看到的清晰 4 倍"。

"increase"的相关固定搭配也常用于描述倍数增长的情况,如:

(a) increase to N times;

(b) increase N times/N-fold;

(c) increase by N times;

(d) increase by a factor of N;

(e) There is a N-fold increase/growth.

分数、百分数、基数中 to 和 by 意思不一样,但倍数增加与减少中"to/by/-"的意思没有差别。以上句型均应表示"增加到 N 倍"或"增加 $N-1$ 倍",且汉语较少用"增加到 N 倍",多译为"增加 $N-1$ 倍"。其中,"increase"常可用"raise, grow, go/step up, multiply"替代,句子含义不变。

(3) The production of integrated circuits has been increased by three times as compared with last year.

译文一 集成电路的产量比去年增加了 3 倍。

译文二 集成电路的产量比去年增加了 2 倍。

(4) The sea freight transport is expected to increase by 7 times and inland transport, 1.6 times.

译文一 海上货物运输预计将增加7倍,内陆运输增加到1.6倍。

译文二 海上货物运输预计将增加6倍,内陆运输增加到1.6倍。

分析 例(3)中"increase by three times"等于"increase three times"或"increase to three times",没有"到"字或"至"字时,需要减1,即译文二为正确译法。例(4)前半部分同例(3),后半部分中"增加到1.6倍"等于"增加0.6倍",但这个数字较小,译为"增加到1.6倍"更符合读者心理期待,便于理解。

二 倍数减少

在汉语中数值增加常用倍数表示,但减少常用分数来表示。常见的描述倍数减少的固定句型如下:

(a) A is *N* times as small (light, slow, ...) as B;

(b) A is *N* times smaller (lighter, slower, ...) than B.

以上两个句型均应译为"A 的大小(重量,速度,……)是 B 的 $1/N$",或者"A 比 B 小(轻,慢,……)$(N-1)/N$"。

(1) A polished screw is twice thinner than an ordinary screw.

译文 抛光后的螺丝比普通螺丝细一半。

(2) In other words, it reflected 10 times less light than all other superblack materials, including Vantablack.

译文一 换句话说,它反射的光比其他所有超黑材料(包括 Vantablack 材料)少10倍。

译文二 换句话说,它反射的光仅有其他所有超黑材料(包括 Vantablack 材料)的十分之一。

分析 例(1)译为"比……细一半"即是"比普通螺丝细二分之一"的意思。汉语中表示倍数减少通常用分数表示,例(2)的译文一中"少10倍"无法直观呈现数值减少的情况,可能导致读者对数值信息的错误解读。

相对于"increase",其反义词"decrease"常用于描写倍数减少的情况,固定句型如:

(a) decrease *N* times/*N*-fold;

(b) decrease by *N* times;

(c) decrease by a factor of *N*;

(d) There is a *N*-fold decrease/reduction.

以上四种表达均可译为"减少至 $1/N$",或"减少 $(N-1)/N$",不应译为"减少 *N* 倍"。其中"decrease"常与"reduce, shorten, slow down, cut"等词替换。

(3) The weight of the electronic device has decreased 4 times.

译文一 电子设备的重量减少了4倍。

译文二 电子设备的重量减少到了原来的四分之一(减少了四分之三)。

分析 在汉语中倍数的减少通常用“几分之几”来表示,而不用“几倍”来描述,因此译文二译法正确。

(4) The genetically modified mosquitoes can drastically reduce the spread of Zika as well as dengue, chikungunya and yellow fever, and the number of deaths is expected to decrease by 5 times.

译文 这些转基因蚊子可以大幅减少寨卡病毒以及登革热、基孔肯雅热和黄热病的传播,且有望将死亡人数减少五分之四。

分析 此处将“decrease by 5 times”译为“减少$(N-1)/N$”,即“减少五分之四”,既保留了汉语中对数值较小的英文分数用汉语数字表示的习惯,又在一定程度上突显了该项生物技术在减少死亡人数方面的重大作用。

第五节 数学运算

科技文体作为一种应用文体,具有客观性和科学性,在翻译此类文本时还需正确处理其中数学运算的翻译问题。

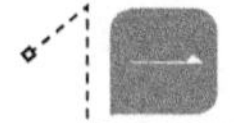

一 加法

“加”在英文中用“plus,and,add”表示;“等于”用“is,make,equal”等词表示。

5+17=22,用英文可表述为:

Five plus seventeen is twenty-two.

Five and seventeen (is) equal (to) twenty-two.

Five and seventeen makes twenty-two.

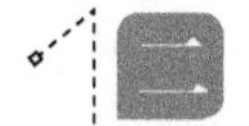

二 减法

“减”用“minus”“take from”或“subtract from”表示,“等于”除了“is”“(is) equal (to)”等,还可以用“leave”表示。

16-6=10,用英文可表述为:

Sixteen minus six is/leaves ten.

Take six from sixteen and the remainder is ten.

Six (taken/subtracted) from sixteen is ten.

三 乘法

“乘”用“times”或“multiplied by”表示。

$2\times6=12$,用英文可表述为:

Two times six is twelve.

Two multiplied by six makes twelve.

四 除法

“除”用“divided by”“into”或“the ratio of…to…”表示。

$56\div7=8$,用英文可表述为:

Fifty-six divided by seven is eight.

Seven into fifty-six goes eight.

$18:6=3$,用英文可表述为:

The ratio of eighteen to six is three.

五 次方

表示“n 次方”的说法:指数采用序数词,底数则采用基数词(4 次方以上适用)。

8 的 2 次方:Eight squared is sixty-four;

3 的 3 次方:Three cubed is nine;

10 的 6 次方:the sixth power of ten(ten to the sixth power);

2 的 9 次方:the ninth power of two(two to the ninth power)。

六 根号运算

根号是用来表示对一个数或一个代数式进行开方运算的符号,写作“$\sqrt[n]{\ }$”。若一个数的 n 次方($n>1$ 且为整数)等于 a,那么这个数叫作 a 的 n 次方根。

平方根:square root;

立方根:cubic root;

11 次方根: eleventh root;

$\sqrt{X} = Y$:The square root of X is Y;

$\sqrt[3]{X} = Y$:The cubic root of X is Y;

$\sqrt[13]{X} = Y$:The thirteenth root of X is Y。

练习题

一、英译汉

1. Incandescent bulbs convert only 10 percent of the energy they draw into light.

2. The thyroid tablets developed by the lab can increase metabolic rates by three times.

3. But sunspot numbers are running at less than half those seen during cycle peaks in the 20th century.

4. The app, known as WalkSafe, can detect cars moving 30 miles per hour at more than 160 feet.

5. As for WASP-12b, there was more than twice the concentration of carbon as usual and almost 100 times more methane than expected.

6. Stars more massive than eight solar masses become neutron stars; and stars greater than 30 solar masses become black holes.

7. So far, 45 percent of the commands that are transmitted from one subject to another-like "call in helicopter" or "enemy ahead"-are correct.

8. The upgraded 4K digital projectors from Sony Corp. will start showing films at the higher, 48 frames-per-second rate, making images appear crisper and more lifelike than the current 24 fps in use.

9. The National Center on Time and Learning was given more than $620,000 to assess the effectiveness of the bracelets by comparing them with MRI scans, and work out a scale that would pinpoint how engaged a student was in lessons.

10. The project's first step is to develop technologies to cut the cost of deep-space robotic probes to one-tenth to one-hundredth the cost of current space missions, which run hundreds of millions of dollars.

二、汉译英

1. 这只是一个起搏器所需能量的五分之一左右。

2. 生物发出的光是海洋中最大的光源:生活在距海平面1500英尺以下的生物中有90%都会发光。

3. 制作完成的薄膜以丝心蛋白为基质,上面布满了直径只有几百纳米(十亿分之一米)的小袋。

4. 有人认为,一颗名为WASP-12b的系外行星适合生命存在,它距离地球约1200光年,质量为木星的 1.4 倍。

5. 研究人员使用氮离子(氮的一种形式)做了这项测试。他们发现,与正水分子相比,负水分子化学反应速度提高了 25% 左右。

6. 该研究团队克服了这一挑战,他们在一个足球大小的小球体外部安装了近 100 台微型相机,每台相机都配有一个 1400 万像素的传感器。

7. 但随着超级巴士这种五米长、有六个轮子的庞然大物的出现,这一局面可能发生改变。它可以搭载 23 名乘客,行驶速度可达每小时 255 千米。

8. 这款太阳能电动车充电一次即可行驶约 624 千米,太阳照射 1 小时便可为车辆增加约 9 千米的行驶里程。

9. 然而,美国国家航空航天局估计现在银河系约有 1000 颗长度超过 1000 米的小行星和 19500 颗长度超过 100 米的小行星,因此,行星研究所的科学家们正在试图探寻保护地球的办法。

10. 这颗系外行星体积比地球大 22%,质量比地球大 80%,这使得它成了“超级地球”。研究人员估计它的平均温度为 254 摄氏度。

第十章 主动式与被动式的翻译

科技文体注重客观性、科学性、准确性，在讨论事物的发展过程、阐述科学原理时，往往着眼于演绎论证的结果，而鲜少关注动作的执行者。为了突出信息焦点，科技英语多使用被动语态，通过主要信息的前置起到强调作用，还存在一些词暗含被动义。汉语没有语态这一说法，一般称作"被动式"（或"被动句"）和"主动式"（或"主动句"）。在科技汉语中，被动式的使用频率并不高。汉语被动式的标记词，如"被、遭（到/受）、蒙、受、获（得）、得到"等，暗含一定的消极意义，汉语被动式远少于主动式。科技翻译要兼顾两种语言的表述习惯和科技文体的语言特点，灵活地转换语态。

第一节 汉语主动式的翻译

在科技汉语中，即使选择主动式，也鲜少出现"你""我""他"等人称代词作主语的情况。为了拉近与读者之间的距离，通常由表示泛指的"人们""大家""众人"等词语来充当主语。另一种情况则是汉语中独有的一大语言现象，即无主语句，这种语法结构既避免了动作执行者充当主语而喧宾夺主的情况，又排除了因使用被动式而夹杂消极情感的可能。在这两种情况下，汉译英时经常会选择被动句，而这一主动与被动之间的转换可以通过句法手段或词汇手段实现。

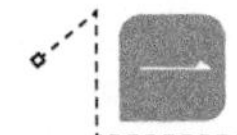

一 句法手段

作为印欧语系的一员，英语是一种屈折语，主要通过动词的形态变化表示被动意义，其被动句的基本句法结构是"be + 过去分词"，并随着时态的变化而变化。正如前文所言，科技汉语多使用主动句并以"人"为主语，彼时多数情况下谓语是表示心理活动的动词，如"想""认为""相信"等。在翻译过程中，往往需要采用句法手段，将主动句转换为被动句，使译文与科技英语的表达习惯相契合。

（1）人们相信这种合金钢是这里能提供的最好的合金钢。

译文一 People believe that this steel alloy is the best available here.

译文二 This steel alloy is believed to be the best available here.

(2)科学家们认为,行星的撞击事件将大量灰尘升至大气层中,遮蔽了阳光数十年,从而导致大规模的气候变化。

译文一 Scientists believe that after the asteroid hit it sent dust up into the atmosphere that blocked the sun for decades.

译文二 It is believed that after the asteroid hit it sent dust up into the atmosphere that blocked the sun for decades.

(3)人们认为在氧气稀薄或缺氧环境下,癌细胞的出现概率会增加。

译文一 People think that a hypoxic, or oxygen-poor environment, can increase cancerous cells.

译文二 It is thought that a hypoxic, or oxygen-poor environment, can increase cancerous cells.

分析 以上三例中,谓语动词均翻译成了相应的心理动词“believe”和“think”,但句法结构发生了变化。原文中,主语“人们”和“科学家”是发出谓语动词的主体,即动作的执行者,在译文中却纷纷隐去,转而强调“相信”和“认为”的内容,也就是实际上的重要信息。

例(1)中,“this steel alloy”作主语,它之所以能成为“这里能提供的最好的合金钢”,并不是自身固有的一种属性,而是通过人们的意志强加给它的,因此选择“is believed to be”的表达来阐释这种被动性。在例(2)和例(3)中,主语既不是形式上原文中的主语,也不是语义上信息的焦点,而是形式主语“it”。一方面,这两句的原文较长,翻译成完整的从句比翻译成短语更符合英语的行文特点;另一方面,“It is believed/thought that”已成为科技英语中表达主体想法的常用结构,是一种约定俗成的翻译方法。

由于汉语语法较为灵活,一个句子即使没有主语也是合乎语法的。而英语则是典型的SV(Subject + Verb)语言,主语是不可或缺的。因此,考虑到这一差异,进行科技文本汉英翻译时一般要选择被动句。

(4)可以预言,钛材在飞机或各种飞行器上的应用将会与日俱增,并且在其他工业中的应用也会扩大。

译文 It can be prophesied that more and more titanium materials will be used in aerospace, and so will they be in other industries.

分析 本例中出现了两次主动式与被动式的转换。第一次出现在谓语动词“预言”的使用中:在译文中,“it”用作形式主语,“prophesy”用被动形式,说明这一预言是由人说出的,而不是自主产生的,但又为了保证科技文本的客观性而将“by people”这一动作执行者隐去。第二次出现在谓语动词“应用”的使用中:原文中的“应用”本来是名词，与前方修饰语部分共同组成带“的”字的偏正短语，充当该句的主语。译文将该部分处理成完整的句子，“钛材”作主语，“在飞机或各种飞行器上”作状语，将名词“应用”转换成动词并用被动结构来表达。从语义上来说，“钛材”只能由人应用在飞机或各种飞行器上。

(5)可根据非水溶性不规则固体的排水量来求得其体积。

译文一 We can find the volume of an irregular solid by the displacement of water, if the solid is not soluble.

译文二 The volume of an irregular solid can be found by the displacement of water, provided the solid is not soluble.

分析 原文是一个由“根据”引导的介词结构，没有主语，两个名词短语分别是“非水溶性不规则固体的排水量”和“体积”。译文一虽然增译了主语“we”，但无论是条件状语从句的处理，还是整个句子，都显得过于口语化。而译文二将“体积”处理为主语，“……排水量”处理为“by”引导的短语，修饰语“非水溶性”处理为条件状语从句，“不规则”则直接译成相应的形容词作定语。连接前后两部分的谓语动词是“find”，“volume”是“find”的结果，“displacement of water”是得出这一结论的根据，因此使用被动语态。同时，以被动结构“can be found”为分界线，前后两个部分长度也较为均衡。

(6)已知人体会产生一定的微弱的磁场。

译文一 It is known that weak magnetic field comes from the human body.

译文二 Weak magnetic field is known to come from the human body.

分析 原文可以划分成两个意群，“已知”和“人体会产生一定的微弱的磁场”，后者是对前者的拓展和补充。译文二用“be known to”来传递“已知”这一语义，此时的“to”做不定式的标识，后接动词短语“come from”，表示“微弱的磁场”是因为“产生于人体”这一状态特性而为人所知的，突出了信息焦点。相比译文一中的“It is known that + 从句”的结构，译文二更加简洁明了。

二 词汇手段

尽管句法手段是英语中表示被动含义的主要手段，但也存在一些词语本身即含有被动意味，尤其是部分名词、介词短语和过去分词性的形容词，它们可以很好地实现语法代偿作用。因此，翻译时也可以直接通过词汇手段实现主被动间的转换。

(1)有种仿生系统，也称“人工神经元”，可以触发捕蝇草的突然闭合。

译文一 A bionic system, or an artificial neuron, can trigger the snap of a Venus fly trap.

译文二 A bio-inspired system, or an artificial neuron, can trigger the snap of a Venus fly trap.

分析 对于原文中"仿生"一词,译文一译成了"bionic",表示体内有电子装置,难以完整地再现原文信息;译文二则译成了复合形容词"bio-inspired"。"仿生"即是说模仿生物的功能和行为来建造技术系统,是一种主动的表达;"bio-inspired"由名词"biology"的缩写加"inspire"的过去分词构成,表示该系统的诞生与生物有关,即因生物激发的灵感而发明这一系统,具有被动的含义。

(2)集成电路在电气工程方面得到了广泛的应用。

译文 The integrated circuit finds wide application in electrical engineering.

分析 原文是主动句,译文从结构上看是主动语态,但实际上表达的是被动意义。谓语动词"find"的施事并非是主语"the integrated circuit",而是文中并未出现的"人们",既"集成电路广泛应用于电气工程"这一事实是由人们发现的;主语、宾语和状语所表达的语义共同构成谓语动词"find"的受事。

(3)中国发展航天事业服从和服务于国家整体发展战略。

译文一 China's space industry obeys and serves the overall national strategy.

译文二 China's space industry is subject to and serves the overall national strategy.

分析 原文的谓语动词是"服从",译文一将其直译成动词"obey",仅表示遵守,不足以传达出"航天事业"与"国家整体发展战略"之间的上下级关系。译文二则用了形容词短语"be subject to"。原文是主动句,"服从"指依顺、听从,主语"中国发展航天事业"是做出依顺这一行为的主体,而宾语"国家整体发展战略"则是它依顺的对象。译文二是被动句,"be subject to"指受到某人或某物的支配,主语"China's space industry"是支配这一动作的承担者,而宾语"the overall national strategy"则是发出该动作的主体。

第二节 被动语态的转换

科技文体注重对客观事实的表达与科学道理的阐释,力求尽可能降低主观性。因此,在科技英语中,被动语态是一种常见的语法现象,业已成为一大语言特点。英语主要是通过动词形态的变化来表达被动意义,从而达成被动语态,而汉语中结构性的被动式较少,且被动式的整体使用频率较低。因此在翻译过程中,科技英语中用被动语态表达的句子,通常要转换成科技汉语中的主动句。

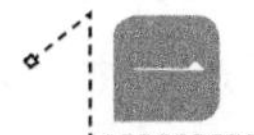

一 无主句

英语为主语显著的语言,主语突出,一般情况下每个句子都有主语;而汉语则是主题显著的语言,主题突出,有的句子即使没有主语也是合乎语法的。翻译科技英语的被动语态时,可以选择无主句的结构,将原文中的主语处理成其他成分。

Gas that comes off the oil later is condensed into paraffin. Last of all the lubricating oils of various grades are produced. What remains is heavy oil that is used as fuel.

译文 随后将从石油分离出来的气体浓缩成煤油。最后产生的是各种等级的润滑油。剩下的便是重油,可以用作燃料。

分析 该例原文由三个被动句组成,在译文中均转换成了主动句。在第一个句子中,"将"是介词,其语义和功能与"把"类似,此处的转换近似于被字句与把字句之间的转换。从结构上来说,原文主语"gas"在译文中充当宾语并前置,原文宾语"paraffin"在译文中充当补语,形成处置式,强调行为结果。从语义上来说,"从石油分离出来的气体"是受事,谓语动词"浓缩"所表示的动作对它施加了影响,使之发生了"浓缩成煤油"这一变化。在第二个句子中,翻译前后受事与谓语动词的位置发生了调换。原文将受事"lubricating oils of various grades"作SV结构中的主语,译文将受事"各种等级的润滑油"作系表结构"……的是……"中的表语,强调产生的结果。在第三个句子中,原文的被动结构"is used as"和译文的谓语动词"用作"都强调了用途。

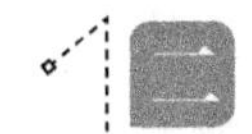

主谓句

(一)保留主语

在大部分英语被动句中,受事充当主语,施事充当宾语,两者由谓语动词的被动态连接,并通过介词"by"引导。有时,在英语中谓语动词表示的行为动作与主语表示的人或物之间是被动关系,而在汉语中却可以通过主动关系表达,这样一来,原文的主语经过翻译后仍然可以扮演主语的角色,但译文的语态已然发生转换。

(1) Many species, most famously the dinosaurs, were annihilated, while many of the survivors underwent radical transformations.

译文 许多物种,甚至包括赫赫有名的恐龙,都在这次事件中灭绝了,而幸存物种们则经受了彻底的演化。

分析 原文的谓语动词"annihilate"表示消灭,是一个及物动词,其所在的被动结构"were annihilated"是过去时态,强调许多物种灭绝这一事实结果,但并未指明施事者。译文的谓语动词"灭绝"表示完全消失或消灭,为不及物动词,"绝"字表示竭、尽,强调消灭的彻底程度,和主语是主动关系,即"许多物种"是发生全球性死亡和消失这一动作的承担者。

(2) When polyphenols CA and CGA were combined with amino acid cysteines found in milk

proteins, their anti-inflammatory effects received a boost.

译文 当多酚 CA 和 CGA 与牛奶蛋白中的半胱氨酸结合,它们的抗炎作用得到了强化。

分析 原文是被动结构“A be combined with B”,如果用主动语态表达,则是“combine A with B”。句中的“polyphenols CA and CGA”即 A,“amino acid cysteines”即 B,它们之所以能结合是因为某个外力的作用,因此是被动的。而在译文中,“结合”指两者之间发生密切联系,这一行为动作是主动的,不要求第三方力的存在,符合中文的表达习惯。

(3) Most known exoplanets have been discovered by indirect means, but planetary scientists believe they can learn a lot more from direct observation.

译文一 大多数已知的系外行星都是被间接方式发现的,但行星科学家相信,他们可以从直接观测中学到更多。

译文二 众所周知,系外行星是通过间接方式发现的,但是行星科学家们相信,通过直接观测会了解到更多。

分析 原文是典型的英语被动式结构,“most known exoplanets”是主语兼受事,“indirect means”是宾语兼施事,“have been discovered”是谓语动词。如果像译文一那样,直译成“系外行星被间接方式发现”,既不符合汉语的语言习惯,又显得生硬刻板。译文二则采用了“……是……的”的结构来体现原文的被动含义,“是”前面的“系外行星”是受事,“是”和“的”之间的“通过间接方式发现”是行为动作,其中“通过间接方式”既是结构上的方式状语,又是逻辑上的施事者。

(二)主宾易位

除了跨语言语态转换外,同一语言内部也可以实现语态的转换。以英语主谓宾结构为例,在转换为被动语态时,原来的宾语变为被动句的主语,而原来的主语在被动句中则由介词“by”引出,此时的主语和宾语交换了位置。这适用于汉语,也适用于英汉翻译。

(1) Polyphenols can be found in many foods, including coffee and tea, fruits and vegetables, red wine, and beer.

译文一 多酚可以被发现于许多食物,包括咖啡、茶、水果、蔬菜、红酒和啤酒。

译文二 咖啡、茶、水果、蔬菜、红酒、啤酒等许多食物中都含有多酚。

分析 原文中,“polyphenols”是主语,“many foods”是宾语,人们能够在后者中发现前者这一元素,但译文一中的“被发现”并不是汉语中的地道表达。译文二中,“许多食物”是主语,“多酚”是宾语,“含有”是谓语动词,直接用主动式表示两者之间的包含关系。

(2) Sensors at road level measure how much light is absorbed by pollutants, and roughly calculate emissions from passing vehicles.

译文一 道路水平的传感器测量多少光被污染物吸收了,并粗略计算出过往车辆的排放量。

译文二 道路上的传感器可以测量污染物吸收了多少光，并大致计算出过往车辆的排放量。

分析 原文中，“light”是主语，“pollutants”是宾语，后者可以对前者施加“absorb”这一动作，使之受到影响从而发生数量上的变化。两个译文均表达出“污染物吸收光”这一命题，区别在于对主动式和被动式的选择。译文一强调受事“光”，译文二强调施事“污染物”，从后半句的动宾结构“计算排放量”可知，本句的信息重点是道路车辆造成的污染，因此译文二更适合，也更符合汉语的表达习惯。

(3) The first X-ray machine for medical diagnosis and disease treatment was invented by the German physicist Russel Reynolds in 1895.

译文一 第一台用于医疗诊断和疾病治疗的X光机被德国科学家拉塞尔·雷诺兹发明于1895年。

译文二 1895年，德国科学家拉塞尔·雷诺兹发明了第一台用于医疗诊断和疾病治疗的X光机。

分析 原文中，“the first X-ray machine”是主语，“the German physicist Russel Reynolds”是宾语，后者发明了前者。在表达“某人发明了某物”这一命题时，汉语通常用主动句（译文二），而不用被动句（译文一）。发明创造往往给人类带来福祉，但在汉语被动句中，“被”字句带有明显的消极色彩，暗含“说话者或听话者不愿该行为动作发生”之意，因此应尽量避免“被发明”这样的表达。

三 增译主语

尽管从语法角度上来看，英语句子必须要有主语，但在翻译过程中，并非所有原文的主语都适合在译文中保留。当原文主语或宾语甚至其他成分都不适合充当译文主语时，往往需要增译主语，“大家”“人们”“我们”等泛指主语尤应注意。

(1) Microplastic pollution has been discovered lodged deep in the lungs of living people for the first time.

译文一 微塑料污染首次被发现于活人肺部深处。

译文二 科学家首次在活人肺部深处发现微塑料污染。

分析 原文主语是“microplastic pollution”，它“存在于活人肺部深处”是客观存在的事实，但并非众所周知。谓语动词“discover”在原文中是被动形式，对应汉语动词“发现”。做句子的谓语时，常用的表达是某人发现某物（译文二），而非某物被某人发现（译文一）。此外，译文二还增译了主语“科学家”，虽然原文中并未出现施事者，但凭常识可知这一事实是由科学家发现的。此外，由于汉语中存在地点主语、时间主语等英语中没有的主语形式，本句还可以不增译“科学家”三个字，译为“在活人肺部深处首次发现了微塑料污染”。

(2) Historically it has been assumed that resistance in disease-causing bacteria is a modern phenomenon driven by clinical use of antibiotics.

译文一 历史上，致病细菌的耐药性被认为是由临床使用抗生素导致的现代现象。

译文二 过去，人们一直认为，致病细菌的耐药性是由临床使用抗生素导致的现代现象。

分析 原文中，“it has been assumed that”是表达观点的常见句式，类似的还有“it is believed /asserted /supposed /proved that...”。如果像译文一那样译成“被认为”，显得过于生硬。译文二增译了泛指主语“人们”，表示“致病细菌的耐药性是由临床使用抗生素导致的现代现象”是大众的观点。

(3) It was not until the 19th century that heat was considered as a form of energy.

译文一 直到十九世纪，热才被考虑成能量的一种形式。

译文二 直到十九世纪，人们才把热视为一种能量。

分析 原文的主语是“heat”，宾语是“a form of energy”，两者之间存在包含关系，这种包含关系是人们赋予的，即人的观点，因此谓语动词“consider”是被动式。译文一保留了原文的主语和语态，直译成“被考虑”；译文二增译了主语“人们”，将“consider”这一动词背后隐含的主语显性化，用“人们才把热视为”的句式引出观点与看法，此举更符合汉语的表达习惯。

近年来，随着简明英语运动的兴起，越来越多的科技工作者开始推崇简洁、质朴的语言。如今，爱思唯尔(Elsevier)、威立(Wiley)、自然(Nature)等众多国际期刊出版社的风格指南及《芝加哥风格手册》(*The Chicago Manual of Style*)、《美国心理学会出版手册》(*Publication Manual of the American Psychological Association*, *APA*)等写作手册都明确建议多使用主动语态(不是全部使用，也不是乱使用)，而《科学》(*Science*)杂志建议适时使用含人称代词的主动语态。译者应根据出版社风格指南尽量做出调整。国内也有学者明确指出：“科技英语采用第一人称和主动语态的文体在英美的确已经成为主流，而且正在对世界各国科技界产生影响。”由此可见，科技英语和被动语态不能画等号，在翻译过程中要充分考虑到读者群体，灵活使用主动语态，使用简洁明了的语言传递信息。

练习题

一、英译汉

1. When the summer sea ice is much reduced, the ocean can absorb and store more heat from sunlight.

2. 3-D-printed creation could be used as a medium to grow algae to produce bioenergy and

also as a tool for studying the coral-algae symbiosis.

3. Tides are caused by the moon and the sun pulling sea water toward them.

4. Air pollution particles are already known to enter the body and cause millions of deaths a year.

5. The dinosaurs weren't killed off with one single blow, but died off gradually over the course of millions of years thanks to competition from mammals.

6. These resistance genes have been spotted not only in folks with Staph infections, but in livestock, like pigs and cattle.

7. The traps are attached to something that is along the animal's regular path, like a tree that the puma has territorially scraped.

8. When motion is detected, the trap gets triggered, resulting in a snapshot of the mountain lion as it strolls by.

9. These cameras even have an infrared flash so that nighttime photos are captured without disturbing the animal.

10. Within a few decades, global temperatures are expected to climb to 1.5 degrees Celsius above pre-industrial levels.

二、汉译英

1. 计算机可分为模拟计算机和数字计算机两种。

2. 新型晶体管的开关时间缩短了三分之二。

3. 一些元素和化合物可以从海水中直接提取出来。

4. 针法是把毫针按一定穴位刺入患者体内,用捻、提等手法来治疗疾病。

5. 物理学中的一个引人注目的部分统称现代物理学,包括电子学、光电学、X射线、放射学等。

6. 产品能值由加工过程中所用的能源、原材料、劳动力,以及能源和原材料的运输所能供的能值组成。

7. 在今后的几个世纪里,大气污染和温室效应不仅使空气变热,还会加热海洋的表层。

8. 人脸识别技术很可能会带来好处,但这些好处需要根据风险进行评估,为此它需要得到适当和细致的监管。

9. 当具有相反磁极的不同区域碰在一起时,压抑已久的惊人数量的磁能就会变成热能释放出来。

10. 人们认为,甲氧西林耐药性与处方药有关,部分原因是在甲氧西林临床使用一年后,英国医院首次分离出了具有甲氧西林耐药性的细菌。

第十一章 从句的翻译

在现代英语的语法中，从句指复合句中不能独立成句，但具有主语部分和谓语部分等，由“that，who，whom，when，why，where，how，whether，which”等引导词（connective）引导的非主句部分。在英文科技文本中结构复杂的长句较多，相较于词汇翻译而言，掌握科技英语较长的从句翻译技巧更为重要。本章将以定语从句和状语从句为主要内容，根据不同的从句类型举例分析其翻译方法。

第一节 定语从句的翻译

定语从句（也称关系从句、形容词性从句），是指一类由关系词引导的从句，因为这类从句的句法功能是做定语，所以人们曾将其称为定语从句。在英语丰富的句子类型中，定语从句最为复杂多变，因此其翻译方法和技巧也是非常丰富的。其原因在于英语和汉语在表达方式和习惯方面的差异较大，而这种差异在科技文本中体现得更为明显。定语从句可分为限制性定语从句和非限制性定语从句，本小节将从这两类定语从句入手，举例分析科技文本中定语从句的翻译方法。

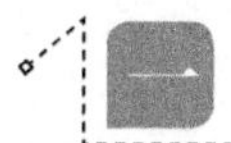

一 限制性定语从句的翻译

限制性关系从句起限定作用，修饰特定的名词或名词短语。从语义上看，限制性关系从句主要起限定作用，修饰特定的人或事物，如果去掉限制性定语从句，整个句子表意会不完整甚至不通顺。因此，在翻译科技文本中的限制性定语从句时，可供我们采用的有顺译法、逆译法、分译法、综合译法、转译法等。

（一）顺译法

顺译法，是指按原句的顺序，把整个句子分割成若干个意义单位或信息单位并逐一译出，再用增译、省译等手段把这些单位自然衔接，形成完整的句意。当限制性定语从句的语序，如因果顺序、先后顺序等与汉语表达基本一致时，可以选用顺译法翻译该从句，以追求在结构和句意上与原文的对等。

(1) Two other techniques, called gasification and oxy-combustion, work by reacting coal with pure oxygen rather than air, and thus produce exhausts that require little treatment before burial.

译文一 其他两种叫作气化和富氧燃烧的技术,用纯氧气而不是用空气烧煤,这样能产生在掩埋之前几乎不用处理的废气。

译文二 其他两种叫作气化和富氧燃烧的技术,用纯氧气而不是用空气烧煤,这样产生的废气在掩埋之前几乎无须处理。

分析 原文是一个复合句。在句子的末尾出现了由关系代词"that"引导的定语从句,形容修饰并限定先行词"exhausts"。译文一采用逆译法翻译,将从句部分前置,译为"能产生在掩埋之前几乎不用处理的废气",比较生硬。译文二则采取顺译法,根据原文句子成分顺序将定语从句"produce exhausts that require little treatment before burial"中的从句部分后置,顺译为"产生的废气在掩埋之前几乎无须处理",承接上文,译文通顺流畅,表达出的句子含义和结构与原文对应。

(2) A system is a collection of hardware, software, data, and procedural components that work together to accomplish an objective.

译文一 系统包括协同完成目标的硬件、软件、数据和程序部件。

译文二 系统包括硬件、软件、数据和程序部件,它们协同来完成目标。

分析 原句为关系代词"that"引导的限制性定语从句,修饰"that"之前所涉及系统包括的各个部分。译文一采用逆译法,将原文中后半部分的从句前置,译为"协同完成目标的硬件、软件、数据和程序部件",句意大体上可以理解,但译文读者有可能认为"系统完成目标的"只修饰"硬件",由此产生歧义。译文二则采取顺译法,按照句子的原本顺序和结构加以处理,用代词"它们"来代替了关系代词"that",将从句顺译为"它们协同来完成目标",使得译文句子顺序合理,指代明确,句意的表达更为清楚流畅。

(3) The electricity is changed into the radio-frequency power which is then sent out in form of radio waves.

译文一 电被转变成接着以无线电波的形式发射出去的射频功率。

译文二 电转变成射频功率后,以无线电波的形式发射出去。

分析 原句为关系代词"which"引导的限制性定语从句,先行词为"radio-frequency power"。译文一采用逆译法,将原句中后半部分的定语从句前置,译为"以无线电波的形式发射出去的射频功率",冗长、拗口,难以将原句蕴含的逻辑关系表达清楚。译文二则采用顺译法,将从句部分译为"以无线电波的形式发射出去",将关系代词"which"省译,使用标点符号逗号来承接上文,使得句子结构完整,逻辑清晰,句意明确。

(二)逆译法

逆译法,是指在翻译时将原本英文原句前的信息后置,而原本在英文原句后的信息则需要前置。使用逆译法翻译的句子结构更加符合汉语表达顺序,因此在遇到使用顺译法译出的句子指向不明确、意义模糊以及缺乏逻辑时,可以考虑使用逆译法,使译文在不影响原文含义的情况下更加符合汉语的表达习惯。

(1) Of 600 seeds that had undergone this experiment, 250 thrived and produced healthy seeds.

译文一 对600粒种子进行了上述实验,有250粒成活并产出了健康的种子。

译文二 上述实验用的600粒种子中,有250粒成活并产出了健康的种子。

分析 原句为关系代词"that"引导的限制性定语从句,修饰的先行词为"600 seeds"。译文一采取顺译法,对关系代词"that"作出省译处理,译为"对600粒种子进行了上述实验",与原文想要表达的含义有出入,也不能表达出"600粒参与实验的种子"与下文"250粒成活种子"之间的包含关系。译文二采用逆译法,将关系代词"that"后面的成分前置,译为"上述实验用的600粒种子中",直观清晰地表达出原句前后成分之间的关系,也更加符合汉语的常见表达习惯。

(2) The particles move faster in the place where the body is being heated.

译文一 粒子在这样的地方运动更快,物体受热部位。

译文二 物体受热部位,粒子运动得较快。

分析 原句是由关系副词"where"引导的限制性定语从句,修饰的先行词为"the place"。本句中"the place"是一个抽象名词,因此使用一个定语从句将其修饰说明。译文一使用直译法,将原文按照句子顺序逐字翻译,译文中出现"这样的地方"与后面的"物体受热部位"所指含义相同,译文冗余复杂,意义表达不明确。译文二采用逆译法,将关系副词"where"后面部分前置,并将关系副词"where"省译,译为"物体受热部位,粒子运动得较快",符合汉语表达方式,句子前后逻辑清晰流畅,译文含义符合原文。

(3) The virus (HIV) invades healthy cells, including white blood cells that are part of our defense system against disease.

译文一 这种病毒(HIV)侵入健康细胞,包括白血球细胞,它们组成部分疾病抵御系统。

译文二 HIV病毒侵入健康细胞,包括组成部分疾病抵御系统的白血球细胞。

分析 原文是由关系代词"that"引导的限制性定语从句,修饰的先行词为"white blood cells"。译文一采用顺译法,按照原文句子顺序逐字翻译,将关系代词"that"增译为代词"它们",句意表达大体上和原文一致,但是译文比较松散,信息传达没有重点。译文二对此定语从句采用逆译法，将关系代词“that”后的句子成分前置，译为“组成部分疾病抵御系统的白血球细胞”，准确清晰地表达出关系代词前后句子成分之间的所属关系，译文突出重点，准确清晰。

(三)分译法

分译法，是指在翻译时将英文中的长句化整为零。在使用分译法翻译限制性定语从句时，可以将原句中的关系代词、关系副词、主谓连接处、并列或转折连接处、后续成分与主体的连接处，按意群将句子切分开，译成汉语分句或独立句，使得句子各个部分之间意义连接紧凑，表达顺畅。

(1)The point near the earth's center, toward which all bodies are drawn, is called the center of gravity.

译文一 靠近地球中心的点，所有的物体都朝着的这个点，被称为重心。

译文二 所有物体都受地心引力牵引，近地球中心之点称为引力中心。

分析 原句包含由“toward + which”引导的定语从句，可划分为三个部分，各部分之间的句意紧密相连。译文一使用顺译法，将原句按照顺序翻译，译文的总体含义与原句表达相同，但逻辑关系有待明晰。译文二使用分译法，将原句中由“toward + which”引导的限制性定语从句的翻译融合于主句中，译为“所有物体都受地心引力牵引”，使得译文逻辑更加清晰，符合汉语表达习惯。

(2)Skyscrapers, factories, and other large buildings now have structural parts of aluminum where many older structures had stone, brick, steel, and other heavier substances.

译文一 摩天大楼、工厂和其他大型建筑现在都有铝结构部件，而许多旧的建筑都有石头、砖、钢和其他较重的物质。

译文二 现在摩天大楼、工厂以及其他大型建筑物中都有铝制构件，而在许多古老的建筑物中，这种构件使用的是石块、砖、钢及其他更为笨重的材料。

分析 原句是由关系副词“where”引导的限制性定语从句，修饰的先行词为“parts of aluminum”。译文一使用部分逆译法，将“many older structures”前置翻译，但其翻译出从句的后半部分“其他较重的物质”指代不明，没能与主句的含义结合。译文二使用分译法，将从句按照主句的逻辑关系拆译成两个小句，指向明确，句意清晰。

(四)综合译法

综合译法，是指在翻译一些像限制性定语从句这样的英语长句时，使用顺译法或逆译法都感到译文表述不正确、不通顺，使用分译法也有困难时，需要仔细推敲原文，或按时间顺序，或按逻辑顺序，有顺有逆、有主有次地对全句加以综合处理。

(1)Scientists are usually engaged in studies that are not aimed directly at the solution of immediate practical problems.

译文一 科学家通常从事不直接针对解决要紧的实际问题的研究。

译文二 通常,科学家们所从事的研究并不针对迫切的实际问题。

分析 原文是由关系代词“that”引导的限制性定语从句,先行词为“studies”。译文一使用顺译法,译文二采用逆译法,将原句中位于后半部分的定语从句前置,译为“通常从事不直接针对解决要紧的实际问题的研究”。译文一受到原句英文思维的影响,没有体现出原句强调的重点。译文二则采用综合译法,全面考虑原句中各个成分之间的语义逻辑关系,先采用顺译法将原句按照本来的顺序翻译,将从句部分与主句的逻辑关系厘清。同时,将“engaged in studies”变通地译为“所从事的研究”,突出原句强调的重点“studies”,让译文更加通顺流畅。

(2) All of the principal units of the lathe are mounted on a bed having ways along which the carriage and tailstock travel.

译文一 车床的所有主要部件都安装在一个床身上,床身上有运载工具和轨道的运行路径。

译文二 车床的主要部件均装在导轨床身上,刀架与尾架可沿导轨滑动。

分析 原句是由“along + which”引导的限制性定语从句。译文一采取分译法,将原文一个完整的句子,拆译成两个独立的分句,但是在分译时忽略了主句中“车床的所有主要部件”与从句中“工具和轨道的运行路径”之间的句意衔接和逻辑关系,使得译文前后分裂,容易让读者产生误解。译文二采用综合译法,综合考虑主句与从句之间的句意衔接和逻辑关系,将从句中“the carriage and tailstock travel”具体化,译为“刀架与尾架可沿导轨滑动”,与译文一相比更加准确流畅。

(3) Mathematical and data-processing techniques that employ digital computers have entered more aspects of our lives than most vigorous proponents dreamt of ten years ago.

译文一 使用数字计算机的数学和数据处理技术已经进入我们生活的方方面面是许多赞助者多年前经常梦想的。

译文二 数字计算机的数学技术与数据处理技术在人们生活中得到广泛应用,这在10年前连最热心的赞助者都不曾梦想过。

分析 原句是由关系代词“that”引导的限制性定语从句,先行词为“Mathematical and data-processing techniques”。译文一采用直译法,将原文按照词语顺序和结构顺序译成一个长句。译文过长,没有停顿,使得译文含义模糊,信息传达不准确,容易产生歧义。译文二将原句的一整个长句根据句意拆分成三个部分,将主句和从句之间的关系通过汉语短句“在人们生活中得到广泛应用”衔接,代词“这”体现完整,使得译文通顺流畅,准确传达原文句意。

（五）转译法

转译法，是指在翻译过程中由于两种语言在语法和习惯表达上的差异，在保证原文意思不变的情况下，译文必须改变词类或对从句进行灵活转换，以使译文符合目的语的表述方式、方法和习惯，而对原句中的词类、句型和语态等进行转换。转译法在翻译限制性定语从句时，为使译文通顺准确，需要发挥汉语的优势，对句子成分进行创造性转译。

（1）A scientist uses many tools for measurement. Then the measurements are used to make mathematical calculations that may test his investigations.

译文一 科学家用许多手段进行测量，然后根据测量结果进行检验自己的研究工作的科学计算。

译文二 科学家采用诸多工具进行测量，然后对所测数据进行数学计算，以检验自己的研究工作。

分析 原句是由关系代词“that”引导的限制性定语从句，先行词为“mathematical calculations”。译文一采用逆译法，将原文后半部分的定语从句前置，译为“检验自己的研究工作的科学计算”，与前文衔接不流畅，译文冗长复杂，没有突出重点翻译对象。译文二采用转译法，将从句“that may test his investigations”转译为“以检测自己的研究工作”。译文二清晰地表明了主句发生的目的。因此，在翻译此类科技文本时，可以采用转译法将定语从句转译为目的状语从句，以便清晰准确地译出主句与从句之间的关系。

（2）Filtration is a simple process of passing the liquid through a sieve in which the holes are too small to allow the passage of the solid.

译文一 过滤是使液体通过筛子的简单过程，筛子上的孔太小，不允许固体通过。

译文二 过滤是使液体通过筛子的简单过程，因筛子孔特别小，固体物不易通过。

分析 原句是由介词“in + which”引导的限制性定语从句，先行词为“a sieve”。通常情况下，在定语从句中“in + which = where”，表示从句发生的地点或场所。但在此句中从句的含义是“筛子孔特别小，固体物不易通过”，与主句之间是因果关系。译文一采用顺译法，将原句按顺序翻译，得到的译文句意大体完整，但没有表达出主句与从句之间的逻辑关系，译文较松散。译文二则使用转译法，将限制性定语从句转译为原因状语从句，译为“因筛子孔特别小，固体物不易通过”准确表达主句和从句之间的逻辑关系，且符合汉语表达习惯。

（3）Continuous cropping is a method of farming in which fields are not given a fallow period between crops.

译文一 连作是一种耕作方法，在这种方法中，农作物之间不存在休耕期。

译文二 连作是一种耕作法，即在两次收成之间不让土地有休耕期。

分析 与例（2）相同，原句也是由“in + which”引导的限制性定语从句，先行词为“a

method of farming”。译文一采用顺译法和增译法，在将原句按照顺序翻译的基础上，增加对先行词“a method of farming”的解释“在这种方法中”，译文冗余复杂，不符合科技文本简洁清晰的特点。译文二考虑到原句中从句的含义是“在两次收成之间不让土地有休耕期”，进一步解释说明主句中提到的“耕作法”。因此，本句可以采用转译法，转译为同位语从句，将从句解释说明的部分更加清晰地翻译出来，使得译文结构清晰，内容完整。

非限制性定语从句的翻译

在语义层面，与限制性定语从句不同的是，非限制性定语从句主要对句中先行词进行补充说明，去掉从句也不会影响对原文含义的理解。在结构层面，在非限制性定语从句的前面往往有逗号隔开，如若将非限制性定语从句放在句子中间，其前后也都需要用逗号隔开。非限制性定语从句一般由关系代词“which，who，whom”或关系副词“when，where”等引导。由于它和限制性定语从句有类似之处，翻译时也可采用一些较为类似的翻译方法，如顺译法、逆译法、分译法、综合译法、转译法等。上述方法在本章第一节已经介绍完毕，这里只借用例句说明几种翻译方法在非限制性定语从句的翻译中的使用。

(1) More often, it consisted of several different units, analyzer, equalizer, display, etc., which had to be interconnected prior to use.

译文 多数情况下，它包括一些不同单元，即分析器、均衡器、显示器等，这些单元在使用前必须互相连接起来。

分析 原句是由关系代词“which”引导的非限制性定语从句，先行词为主句中所提到的“分析器、均衡器、显示器等不同的单元”。若将此非限制性定语从句去掉也不影响原文含义，该从句起到对主句成分的进一步补充说明的作用。根据原句的结构特点，结合上一节中的翻译方法，不难发现，本句使用顺译法并将关系代词“which”增译为代词“这些”承接上文，即可将原文的句意表达准确，译文流畅无误。

(2) For pyrite and pyrrhotite concentrates, which have had limited commercial value to date, the desulphurization process offers an economic means of producing both elemental sulphur and high-grade ironoxide.

译文一 对于黄铁矿和磁黄铁矿，它们目前商业价值有限，脱硫工艺提供了生产元素硫和高品位氧化铁的经济手段。

译文二 对于至今工业价值仍有限的黄铁矿及磁黄铁矿而言，这种脱硫法为同时生产元素硫和高级氧化铁提供了一种经济实用的手段。

译文三 黄铁矿及磁黄铁矿的工业价值至今有限，而该脱硫法为用这两种铁矿生产元素硫和高纯氧化铁提供了经济实用的手段。

分析 原句是由关系代词“which”引导的非限制性定语从句，先行词为主句中所提到的“pyrite and pyrrhotite concentrates”。原文中使用非限制性定语从句的目的是补充说明，定语部分的意思是“黄铁矿及磁黄铁矿的工业价值至今有限。”译文一采用顺译法，按原句的句子结构和顺序翻译，增译代词“它们”看似有所指，但未将非限制性定语从句对主句补充说明的内容与主句恰当的衔接。译文二将从句部分前置翻译成限定成分，译为“对于至今工业价值仍有限的黄铁矿及磁黄铁矿”，会使译文读者认为有些黄铁矿和磁黄铁矿价值高，有些不高，从而造成歧义。译文三考虑到汉语表达的特点，采用综合法将原文译为“黄铁矿及磁黄铁矿的工业价值至今有限”，准确无误地体现出“黄铁矿及磁黄铁矿”的特点，重点突出，译文通顺流畅，表达准确无误。

(3) One of the greatest promoters of structural organic chemistry around the turn of the century was Emil Fisher, who, as early as 1893, had already the structure of cellulose as a polysaccharide in mind.

译文一 在19世纪末和20世纪初，结构有机化学最伟大的推动者之一是埃米尔·费希尔，早在1893年，他就已经认为纤维素的结构是聚糖。

译文二 埃米尔·费希尔是20世纪前后结构有机化学领域的佼佼者。他早在1893年就想到了纤维素的结构是聚糖。

分析 原句是由关系代词“who”引导的非限制性定语从句，先行词为主句中所提到的化学家Emil Fisher(埃米尔·费希尔)。很显然，原句中的非限制性定语从句是用来进一步介绍埃米尔·费希尔的相关信息。译文一采用顺译法，按照原句的标点符号划分和结构顺序进行翻译。尽管句意表达基本完整，但译文较为松散，不符合科技文本简洁的特点。与英语不同，汉语中状语可以放在主语之后。因此，译文二将一个长句按照句意分译为两个句子，第一句介绍埃米尔·费希尔的基本信息，第二句介绍了原句的主要强调点“埃米尔·费希尔”本人，使得译文主次更明确，结构更清晰，信息传达更完整。此外，译文二还用“佼佼者”一词替代“最……之一”这一典型的欧化句式，使译文更简洁地道。

(4) A chrome-plated surface, such as that with which we are familiar in automobile parts and plumbing fixtures, takes a high polish that is not easily tarnished or scratched.

译文 例如我们在汽车和卫星设备上经常看到的那种镀铬表面，其表面光亮度很高，不易变色或刮伤。

分析 原句是一个复合句，其中有一个由“with + which”引导的非限制性定语从句以及一个由关系代词“that”引导的限制性定语从句。对于此类复合句的翻译，在上一节中也有过分析，最好使用综合译法，厘清原句的各个成分和结构之间的逻辑关系和表达含义,而不是单纯使用一种固定的翻译方法。在此句中,第一个非限制性定语从句使用逆译法,译为先行词“镀铬表面”的定语,又使用顺译法处理“镀铬表面”的特质。在两个定语从句的衔接处又使用分译法将原文的三个小句融合成两个句子,使得句意衔接紧凑,表达明确。

(5) A Chinese delegation have been sent to European countries, who will negotiate trade agreements with the respective government.

译文一 中国代表团已被派往欧洲国家,他们将与各自的政府谈判贸易协议。

译文二 中国已派出代表团前往欧洲各国,以便与各国政府谈判贸易协定。

分析 原句是由关系代词“who”引导的非限制性定语从句,对主句中的“Chinese delegation”做补充说明。译文一使用顺译法,同时将关系代词“who”增译为“他们”,是顺译法比较常见的翻译模式。但是不难发现,译文一的前后两个小句互相独立存在,在句意上没有逻辑关联,不能准确表达原句含义。译文二采用转译法,分析出从句的句意是表示目的,即中国代表团前往欧洲各国的目的是与各国政府谈判贸易协定,因此将非限制定语从句转译为目的状语从句,更加清晰直观地传达原文之意。

(6) To find the pressure we divide the force by the area on which it presses, which gives us the force per unit area.

译文一 为了求压力,我们将力除以它所压的面积,这就得到了单位面积的力。

译文二 力除以它所作用的面积,得出单位面积受到的力,就可以求出压强。

分析 原句中有两个定语从句,一个是由介词“on + which”引导的限制性定语从句来修饰先行词“the area”,另一个由关系代词“which”引导的非限制性定语从句来解释或说明前文的主句。译文一采用顺译法,按原句顺序翻译,句意传达基本完整,但是译文小句较多,比较分散,不能准确表达原文各个部分之间的逻辑关系。译文二就考虑到了逻辑关系这一重要因素,将两个定语从句转译为结果状语从句,使用动词词组“得出”和“求出”,直截了当地表明主句发生的结果,逻辑清晰,表达流畅。

(7) Electronic computers, which have many advantages, cannot carry out creative work and replace man.

译文一 有许多优点的电子计算机不能完成创造性工作并取代人。

译文二 电子计算机尽管优点诸多,但不能完成创造性工作,也不能代替人。

分析 原句是由关系代词“which”引导的非限制性定语从句,先行词为“electronic computers”。译文一采取逆译法,将非限制性定语从句前置,修饰先行词,译为“有许多优点的电子计算机”。同时,译文一还将原文的三个小句融合为一个句子,句意表达基本正确,但缺少逻辑关系,不符合汉语表达习惯。译文二采用转译法,根据原句各小句的句意分析出蕴含的逻辑关系，将非限制性定语从句转译为让步状语从句，用“尽管……但(是)……”的句子结构表明句意，更加符合汉语表达习惯。

第二节　状语从句的翻译

状语从句是指句子用作状语时，起副词作用的句子。状语从句可以修饰谓语、非谓语动词、定语、状语或整个句子。状语从句根据其作用可分为时间、地点、原因、条件、目的、结果、让步、方式和比较等从句，一般由从属连词引导，也可以由短语引导。作为英语复合句中很重要的一种句型，状语从句与定语从句的翻译方法基本相同，可采用顺译法、逆译法、分译法、综合译法、转译法等。本小节根据状语从句的类型，分别对科技文本中的条件状语从句、时间状语从句、原因状语从句、结果状语从句、让步状语从句以及目的状语从句的翻译举例说明。

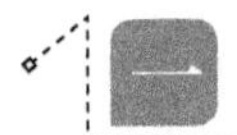

一　条件状语从句的翻译

条件状语从句，是指在复合句中表示主句动作发生条件的状语从句，一般由“if，unless”等连词引导。科技文本中，条件状语从句时可采用的翻译方法有顺译法、逆译法、综合译法等。

(1) If the world is to continue burning fossil fuels while avoiding the consequences, then it will need a lot of CCS(Carbon Capture and Storage).

译文　如果人类要继续燃烧矿物燃料，且免除其不利影响，将会大量用到碳捕捉与储藏技术。

分析　原句是由“if”引导的条件状语从句，从句部分是下文“it will need a lot of CCS”的条件。根据原句结构顺序可采取顺译法，对于科技文本中此类由从属连词引导并位于句首的条件状语从句，不妨依照句意考虑使用顺译法，使译文与原文在结构和语义上对等。

(2) Electricity would be of very little service if we were obliged to depend on the momentary flow.

译文一　如果我们需要依靠瞬时电流，电没有多大用处。

译文二　若非得依赖瞬流时，电的用途甚微。

分析　原句同样是由“if”引导的条件状语从句，我们需要着重分析主句和从句之间的逻辑关系。译文一采用逆译法，将从句前置处理，句意表达基本正确，但缺少深入的逻辑分析。主句“Electricity would be of very little service”与从句“we were obliged to depend on the momentary flow”之间并不只是句子形式层面的条件限制,更有内部的时间联系。因此,译文二采用转译法,将条件状语从句转译为时间状语从句,用“若……时”来表明主句与从句之间的逻辑关系,使得译文更加通顺流畅。

(3) The experiment will become unsustainable unless the growth of microorganisms in utensils is held in check.

译文一 实验将无法继续进行,除非控制器皿中微生物的增长。

译文二 除非控制器皿中的微生物的增长,实验将无法持续。

分析 原句是由“unless”引导的条件状语从句,限制主句“The experiment will become unsustainable”。译文一采用顺译法,按原句顺序翻译,句意表达基本正确完整,但是译文不符合汉语书面表达习惯。一般来讲,汉语中会将条件置于结果之前,先说条件,后谈结果。因此,译文二采用逆译法,将“unless”后的从句前置,使得译文更加符合汉语表达习惯,更加通顺流畅。

(4) The measurement results show, on condition that identical load, vacuum is stabilized in one value.

译文一 测试结果表明:在相同负载的条件下,真空稳定在一个值。

译文二 测试结果表明:相同负荷下,真空度最终稳定在某一值。

分析 原句是由“on condition that”引导的条件状语从句。在结构上原句可分为三个小句,总体结构比较松散。译文一采用顺译法将原句按照句子顺序翻译,译文总体意思表达准确,但是没有考虑到科技文本简洁明了的特点。因此,译文二采用综合译法,将原文三个小句合并译为一个小句,语言表达精炼准确,译文完整通顺。

(5) Supposing that a switched discrete time linear system is globally asymptotically stable, a fuzzy switch method for the system is presented.

译文一 假设切换离散时间线性系统是全局渐近稳定的,提出了该系统的模糊切换方法。

译文二 在已知离散线性开关系统全局渐近稳定的前提下,提出了一种系统的模糊切换方法。

分析 原句是由“supposing that”引导的条件状语从句。译文一采取直译法,直接将“supposing that...”译为“假设……”,基本翻译出了原文想要表达的含义,但与主句之间缺少可以表达衔接性和关联性的词语。此外,译文一中出现语法错误,前半部分是条件状语,其中包含主语,而后半部分缺少主语,意思模糊。因此,译文二将“supposing that...”译为“在……的前提下”放在句首,尽管前半部分没有主语,但译文读者可以分析出是作者(团队)提出了这种方法,表述符合汉语表达习惯,且更加准确完整。

二 时间状语从句的翻译

时间状语从句，是指用表示时间的连词连接一个句子作状语的主从复合句。连接时间状语从句的连接词有“when, before, after, while, as soon as, until, since”等。翻译时间状语从句时同样可以使用顺译法、逆译法、分译法等复合句常用的翻译方法，使得句子中表示时间含义的成分清晰完整，句意通顺。

(1) The test data will change to different degrees when the new system is put into use.

译文一 测试数据将会发生不同程度的变化在使用新系统时。

译文二 在使用新系统时，测试数据一定会发生不同程度的变化。

译文三 测试数据在使用新系统时会发生不同程度的变化。

分析 原句是由“when”引导的时间状语从句，阐明主句“The test data will change to different degrees”的时间条件。译文一使用顺译法，直接按原句顺序翻译，将时间状语从句放在主句之后，译文句意混乱复杂，不符合汉语表达习惯。译文二采用逆译法，将时间条件置于事件发生之前，作为该事件发生的先决条件，译为“在使用新系统时”，看似正确，实则没有强调出原句中的中心词“测试数据”，译文较为松散。译文三则采用综合译法，将时间状语前置的同时，突出强调“测试数据”这一中心词，译为“测试数据在使用新系统时”，更加符合汉语表达习惯，译文信息主次分明，准确通顺。

(2) Production has been suspended while safety checks are carried out.

译文一 生产被暂停当安全检查期间。

译文二 安全检查期间，生产暂停。

分析 原句是由“while”引导的时间状语从句，从句是主句“Production has been suspended”的先决条件。译文一使用顺译法，首先翻译了主句，其次翻译“while”引导的时间状语从句，不难发现该译文不通顺且完全不符合汉语表达习惯。译文二则使用逆译法，将时间状语从句前置翻译为“安全检查期间”，表达准确得当，完全符合汉语表达习惯。同时译文二将主句中的被动式译为主动式“生产暂停”，言简意赅，准确地译出了原句要表达的含义。

(3) The next operation must wait until the drug is discharged from the body.

译文一 下一步操作必须直到受试将药物排出体外才能进行。

译文二 (受试)药物排出体外后，方可实施下一步操作。

分析 原句是由“until”引导的时间状语从句。“until”在时间状语从句中一般译为“直到(……才)”。译文一采用顺译法，将原句按顺序翻译为“直到受试将药物排出体外才能进行”，译文表达不正确，句意模糊。译文二采用分译法，将原句分译为两个小句,使用的句子结构为“……后,方可……”,译文主次分明,逻辑清晰,句意表达准确简洁。

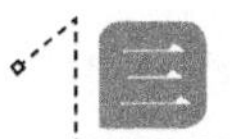

三 原因状语从句的翻译

原因状语从句是指在句中用来说明主句原因的句子,常由“because, as, since, now that”等引导词引导。在科技文本中,翻译原因状语从句时可以采用顺译法、逆译法、综合译法等,使得主句和从句之间的因果关系明确,逻辑结构清晰。

(1) But exploiting these brings technological and regulatory difficulties, and is in any case controversial because it would do damage to deep-ocean ecosystems.

译文一 但使用它们会带来技术和管理上的困难,无论怎样都会有争议,因为它会对深海生态系统造成破坏。

译文二 但使用这些金属结核会带来技术和管理上的困难,并且由于这会对深海生态系统造成破坏,所以无论怎样都会有争议。

分析 原句是由“because”引导的原因状语从句。译文一采用顺译法,将原句按照顺序翻译为“因为它会对深海生态系统造成破坏”,结构欧化,句意不清,有违科技翻译之初衷。译文二采用综合译法,先按照原句顺序翻译,并将原文中使用的代词“these”改译为名词“这些金属结核”;后使用分译法,在原因状语从句的从属连词“因为”前后拆译句子,清晰明确地译出造成“争议”的原因,逻辑清晰,表达流畅。

(2) Since transistors are extremely small in size and require only a small amount of energy, they can make previously large equipment much smaller.

译文一 由于晶体管的体积非常小,耗电较少,它们能使原来庞大的设备缩小许多。

译文二 晶体管的体积极小,耗电较少,因此能将原先庞大的设备大为缩小。

分析 原句是由“since”引导的原因状语从句。译文一采用直译法,将原因状语从句译为“由于晶体管的体积非常小,耗电较少”,若译文后半句添加“所以”或“因此”会更符合汉语语法习惯,句意大体完整清晰。译文二使用顺译法,先强调原因“晶体管的体积极小”,再译出由于这一原因而产生的结果“能将原先庞大的设备大为缩小”,译文符合汉语意合的特点,句意完整,逻辑清晰。

(3) Nicotine, for example, is associated more with rural populations than urban ones, because people living in the countryside are more likely to smoke.

译文一 例如,尼古丁与农村人口的关系比城市人口更密切,因为生活在农村的人更有可能吸烟。

译文二 例如,农村居民受尼古丁的影响比城市居民更大,其原因是在农村生活的人更有可能吸烟。

分析 原句是由"because"引导的原因状语从句,从句说明解释了为什么"尼古丁与农村人口的联系比城市人口更多"。译文一采用顺译法,将原因状语从句译为"因为生活在农村的人更有可能吸烟",从句部分的翻译基本正确,但汉语中较少使用"……(结果),因为……",受英语表达习惯影响,不够通顺,句意模糊。译文二也采用顺译法,但将从句中词语的词性转换为名词,译为"其原因是在农村生活的人更有可能吸烟",更能突显原因,译文结构完整,句意清晰,表达流畅。

四 结果状语从句的翻译

结果状语从句,是指在复合句中用作结果状语的句子,一般放在主句之后,由"so...that"或"such...that"引导。在科技文本中,翻译结果状语从句时,要了解"so...that"或"such...that"的搭配用法,可以采用顺译法、逆译法、综合译法等方法加以处理。"so...that"或"such...that"一般不译作"如此……以至于/以致"句型,应视情况灵活翻译。

(1) Now this machine gives us such high resolution that we can see very small specks of calcium.

译文 现在这台机器分辨率很高,我们通过它能看到钙微粒。

分析 原句是由"such...that"引导的结果状语从句。"such...that"结构在本句中的搭配为"such + adj.(high) + n.(resolution) that"。译文采用分译法将原句拆译为两个小句,首先强调"这台机器分辨率很高",其次说明结果为"我们能通过它看到钙微粒",译文逻辑清晰,符合汉语表达习惯。

(2) We are reproducing ourselves at such a rate that our numbers threaten the ecology of the planet.

译文一 我们的繁衍速度太快,以至于我们的人口数量已经对地球的生态系统造成了威胁。

译文二 人类繁衍速度过快,其数量将危及地球的生态系统。

分析 原句是由"such...that"引导的结果状语从句。"such...that"结构在本句中的搭配为"such + a/an + n.(rate) that"。译文一采用顺译法,将结果状语从句译为"以至于我们的人口数量已经对地球的生态系统造成了威胁",译文受原文影响,"翻译腔"较浓。译文二采用综合译法,在理解原文句意和逻辑关系的基础上,省译连接词"that",并根据文意增译程度副词"过快",译为"繁衍速度过快,其数量将危及地球的生态系统",译文因果关系明确,更加符合汉语意合特点,结构完整,句意通顺。

(3) New findings from researchers at Harvard and elsewhere suggest that a surprising number

of people are face-blind, so bad at recognizing faces that they routinely snub acquaintances and have trouble following movie plots.

译文一 来自哈佛大学及其他地方的研究人员发现,有相当一部分人都患有脸盲症,而且这个数目是惊人的。这些人如此无法辨识脸部容貌,以至于经常对熟人视而不见,看电影时也跟不上情节。

译文二 来自哈佛大学及其他地方的研究人员发现,有相当一部分人都患有脸盲症,而且这个数目是惊人的。这些人完全无法辨识脸部容貌,所以经常对熟人视而不见,看电影时也跟不上情节。

分析 原句较长,可以划分为一个由"that"引导的宾语从句和一个由"so...that"引导的结果状语从句。"so...that"结构在本句中的搭配为"so + adj.(bad) + n.(faces) that"。对于科技文本中较长句子的翻译,一般可以采用综合译法。译文一将结果状语从句译为"如此…… 以至于……"的结构,译文结构死板,内容模糊,"翻译腔"颇浓。译文二使用综合译法,首先使用分译法,将结果状语从句之前的部分拆译为三个短小句,小句之间句意衔接流畅,逻辑清晰;其后,使用顺译法,用"所以"引出"完全无法辨识脸部容貌"会导致的结果,符合汉语因果关系的表达习惯,逻辑关系明确,译文通顺流畅。

(4) Some stars are so far away that their light rays must travel for thousands of years to reach us.

译文 有些恒星十分遥远,其光线要经过数千年才能到达地球。

分析 原句是由"so...that"结构引导的结果状语从句,在本句中的搭配为"so + adj.(far) + adv.(away) that"。译文采用分译法,将原句的整句根据句意和逻辑关系拆分成两部分,首先强调"有些恒星十分遥远",其次略去连接词"that",直接说明结果"其光线要经过数千年才能到达地球",句意表达准确,逻辑清晰。

五 让步状语从句的翻译

让步状语从句,是指在复合句中表示"虽然……""尽管……""即使……"等概念的状语从句,一般由"though, although, while, even if, even though, whether... or ..., despite, in spite of"或"no matter + 疑问词"等引导。在科技文本中,翻译条件状语从句时可以采用顺译法、逆译法、综合译法等方法。汉语状语可以放在主语前,也可以放在主语后,翻译时可以灵活处理。

(1) Though different from each other they have strangely similar properties.

译文一 虽然并不相同,但它们有着不可思议的相似性质。

译文二 虽然它们并不相同,但却有着不可思议的相似性质。

译文三 它们虽然并不相同,但却有着不可思议的相似性质。

分析 原句是由"though"引导的让步状语从句。汉语中一般使用"虽然……但(是)……"的结构表达让步,但英语中只使用"though/although"或者"but"。译文一采用顺译法,将原句按顺序翻译,译文受到英文句式的影响,不符合汉语表达习惯。译文二将让步状语从句译为"虽然……但(是)……"结构,但没有注意到主语的位置。在翻译状语从句时,需要将主语作移位处理,突出强调部分。译文三使用顺译法和分译法结合的综合译法,首先将"though"译为"虽然",表明强调让步关系,其次使用分译法,将原本的整句拆译为两个小句,并将主语移位到句首,清晰地译出让步关系,译文简洁清晰,句意表达完整准确。

(2) Though they are mutually repulsive because they are positively charged, protons are held together by a phenomenon called the strong nuclear force.

译文一 尽管质子之间因带正电荷而相互排斥,但它们通过强核力聚集在一起。

译文二 质子之间虽然因带正电荷而相互排斥,但它们通过强核力聚集在一起。

分析 原句是由"though"引导的让步状语从句。译文一采用顺译法,按照原文的让步关系和顺序直接译为"尽管质子之间因带正电荷而相互排斥",句意准确,表达清晰,但没有将主语进行移位处理。译文二在译文一的基础上,将主语进行前置处理,译文更加通顺准确,符合汉语表达习惯。同时可以看出,在翻译科技文本中的让步状语从句时,若原文将让步关系置于句首,译文通常采用顺译法。

(3) No matter how soft the light is, it still fades carpets and curtains in every room.

译文一 无论多么柔和的光线,也仍然会让所有房间里的地毯和窗帘褪色。

译文二 光线无论多柔和,仍会让每个房间里的地毯和窗帘褪色。

分析 原句是由"no matter how"引导的让步状语从句,同例(2)相同,将表示让步关系的从属连词置于句首。译文一采用顺译法,将原句按顺序译为"无论多么柔和的光线",与下文衔接不流畅,句意表达不够清晰。因此,译文二将主语移位到状语之前,再按原文语序翻译,表明让步关系的同时,将主句部分增译能愿动词"能够",使得句子表达流畅,句意完整。

六 目的状语从句的翻译

目的状语从句,是指在复合句中用以补充说明主句中谓语动词发生的目的的状语从句,一般位于"that, so that, in order that, lest, for fear that, in case (that)"等引导连词之后。在翻译科技文本中的条件状语从句时,可以采用顺译法、逆译法、综合译法等。

(1) In order that his analysis reflected the most up to date demographic information, Dr.

Smith timed the experiment to coincide with a census.

译文 为确保分析报告能够反应最新的人口统计信息,史密斯博士安排实验和人口普查同步开展。

分析 原句是由"In order that…"引导的目的状语从句。通常,由"in order that"开篇的句子可以直接使用顺译法,先将目的强调在句首,再将句子主体部分按照顺序和句子成分直接翻译。因此,译文使用顺译法,先译出目的,再译出事件本身,逻辑清晰,表达准确。

(2) The watt is a small unit of power so that we use the kilowatt instead.

译文 瓦是小功率单位,人们因此采用千瓦。

分析 原句是由"so that"引导的目的状语从句。译文使用分译法,在从属连词"so that"处拆分开来,先说明客观情况"瓦作为功率单位很小",再译出目的"人们因此采用千瓦",译文通顺流畅,句意表达准确完整。

(3) This process is necessary in order that life may run its appointed course.

译文一 这个过程是必要的,目的是让生命自己运行。

译文二 这个过程对生命正常运行不可或缺。

分析 原句是由"in order that"引导的目的状语从句。译文一采用顺译法和分译法,先说明"过程是必要的",再译出"目的是让生命自己运行",译文不通顺,冗余复杂,容易产生歧义。译文二采用转译法,将主句与从句之意整合,用最简单直接的句型"……对……不可或缺"来表达目的,句意完整,译文流畅。

第三节 名词性从句的翻译

名词性从句是指在主句中做名词成分的句子,在句子中的功能相当于名词词组,可以作主语、宾语、表语、同位语、介词宾语等,因此根据它在句中不同的语法功能,名词性从句又可分别称为主语从句、宾语从句、表语从句和同位语从句。科技英语中引导名词性从句的常用从属连词有"that,what,which,whatever,when,where,how,why,if,whether"等。在翻译科技文本中的名词性从句时,可采用顺译法、逆译法、分译法、综合译法、转译法等。本小节将以名词性从句的四大类型为引,举例说明如何翻译科技文本中的主语从句、宾语从句、表语从句和同位语从句。

一 主语从句的翻译

主语从句,是指在复合句中充当主语成分的句子。引导主语从句的从属连词一般有

“that,whether,if”；引导主语从句的连接代词一般有“who,whoever,what,whatever,which,whichever”等；引导主语从句的连接副词一般有“why,when,where,how”等。主语从句可以分为两大类，一种是与普通主语一样位于句首的主语从句；另一种是用“it”作形式主语，把真正的主语从句移到主句后部的主语从句。在翻译科技文本中的主语从句时可以使用顺译法、逆译法、转译法等。

(1) What we call air is a mixture of several different gases.

译文一 我们称之为空气的东西是数种气体的混合物。

译文二 所谓空气，即几种不同气体的混合物。

分析 原句是由连接代词“what”引导的主语从句，句子结构为“主语从句＋系动词＋其他成分”，从句部分“what we call air”用作整个句子的主语。译文一采用顺译法和增译法，将从句译为“我们称之为空气的东西”，其句意虽总体上完整，但受英语句子结构影响较大，具有很明显的“翻译腔”，译文略显冗余复杂。译文二同样使用顺译法，但巧妙地将主语从句部分处理为“形容词＋名词”的形式，译为“所谓空气”，更加符合汉语表达习惯，译文通顺流畅，简洁易懂。

(2) It is widely acknowledged that the computer and other machines have become an indispensable part of our society.

译文一 计算机和其他机器被普遍认为已经成为我们社会必不可少的一部分。

译文二 人们普遍认为，计算机和其他机器已经成为我们社会必不可少的一部分。

分析 原句是由从属连词“that”引导的主语从句，句子的基本结构为“It is ＋过去分词(acknowledged)＋that 引导的主语从句”，“believe, state, know (learn, understand), recommend, report”等动词的过去分词形式常用于此类句型中，一般翻译为“据信、据称、据了解、根据建议、据报道”等。原句中句首的“it”为形式主语，真正的主语是“that”引导的主语从句。译文一使用逆译法，将原句中的从句部分前置，并使用被动语态译为“计算机和其他机器被普遍认为已经成为……”，译文翻译痕迹明显，不符合汉语表达习惯，句意模糊。译文二采用顺译法和增译法，在将原句按照顺序翻译的基础上，增译原文没有出现的行为主体“人们”，使得译文通顺流畅，符合汉语表达习惯。

(3) It is likely that the endosperm of many other plants goes through a similar free-cell, liquid stage.

译文一 可能许多其他植物的胚乳也经历了相似的游离细胞和流体胚乳阶段。

译文二 许多其他植物的胚乳可能也经历了相似的游离细胞和流体胚乳阶段。

译文三 许多其他植物的胚乳也经历相似的游离细胞和流体胚乳阶段，这是有可能的。

分析 原句是由从属连词“that”引导的主语从句，句子的基本结构为“It is＋形容词＋that 引导的主语从句”，其中句首的“It”为形式主语，真正的主语是“that”引导的主语从句。译文一采用顺译法，将原句按照顺序翻译的同时，把主句中的形容词“likely”译为“可能”，置于句首，译文整体受英语句式影响较大，不符合汉语表达习惯。译文二则将“likely”后置，译在主语从句的主语“the endosperm of many other plants”之前，符合汉语表达习惯，译文句意完整，但原句中的重点部分未突出。译文三将逆译法和分译法结合，优先呈现从句中的重点信息，并将原句拆译为两个小句，其中第二个小句为总结性评价，突出强调原句中的重点信息，译文准确通顺，主次分明。

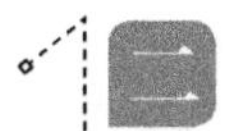

二 宾语从句的翻译

宾语从句是指在复合句中充当宾语成分的句子。宾语从句一般位于及物动词、介词或复合谓语之后，起到宾语的作用。引导宾语从句的从属连词一般有“that（常可省略），whether，if”；引导宾语从句的连接代词一般有“who，whose，what”等；引导宾语从句的连接副词一般有“when，where，how，why”等。顺译法、转译法、综合法等翻译方法在翻译科技文本中的宾语从句时同样适用。

（1）Studies by George and his associates have shown that coconut milk and other liquid endosperms contain a number of different chemical substances.

译文 乔治及同事的研究表明，椰乳和其他的胚乳包含大量不同的化学物质。

分析 原句是由从属连词“that”引导的宾语从句，做主句谓语动词“show”的直接宾语。因此可以直接使用顺译法，将宾语从句置于动词“show”之后，直接说明研究的结果是“椰乳和其他的胚乳包含大量不同的化学物质”，译文准确清晰，通顺流畅。

（2）We consider it of great importance that we should make a thorough study of catalysis.

译文一 我们认为极为重要的是要对催化作用进行彻底研究。

译文二 我们认为对催化作用进行彻底研究是极为重要的。

分析 原句是由从属连词“that”引导的宾语从句，其中主句中谓语动词“consider”后的“it”为句子的形式宾语，真正的宾语是由“that”引导的宾语从句。译文一采用顺译法，先将主句翻译为“我们认为极为重要的是”再将从句翻译为“要对催化作用进行彻底研究”，译文句意表达基本准确，但受英语句式影响较大，不符合汉语表达习惯。译文二采用逆译法，将真正的宾语“we should make a thorough study of catalysis”翻译在谓语动词“认为”后，并在句末处强调宾语部分是“极为重要的”，译文重点突出，主次分明，流畅准确，符合汉语表达习惯。

（3）Some materials will break sharply, without plastic deformation, in what is called a brittle failure.

译文一 有些材料会突然折断，因为没有塑性变形，这被叫作脆断。

译文二 有些材料未经塑性变形便突然断裂,这种现象叫脆性失效。

分析 原句是由介词"in" + 连接代词"what"引导的宾语从句。译文一采用顺译法,按原文顺序翻译,增加连词"因为",试图表明句子之间的逻辑关系,看似无误,实则句意模糊,结构松散,不能准确译出原文的重点信息。译文二采用综合译法,先将原句中"without"引导的复合结构前置翻译为形容词"未经塑性变形的"以修饰主语,说明材料会断裂的原因,再把宾语从句按照顺序翻译,主次分明,逻辑清晰,符合汉语表达习惯。

三 表语从句的翻译

表语从句,是指在复合句中充当表语成分,说明主语"是什么"的句子。表语从句时,可以由名词、形容词、相当于名词或形容词的词或短语充当句子中的表语成分。引导表语从句的从属连词一般有"when, where, why, who, how, that"等。翻译科技文本中的表语从句时,一般采用顺译法和综合译法,按照原文的顺序把句子翻译成汉语,常用的译文句子结构为"……是……"。

(1) A more likely reason for the change, the researchers argue, is that scientists and inventors are producing work based on narrower foundations.

译文一 这一变化更可能的原因是,研究人员认为,科学家和发明家研究的领域更"窄"了。

译文二 研究人员认为,造成这一变化更可能的原因是科学家和发明家研究的领域更"窄"了。

分析 原句是由从属连词"that"引导的表语从句,用来说明主语"A more likely reason for the change"。译文一采用顺译法,将原句所包含的信息基本翻译完整,但是译文受英文句式影响较大,没有顾及主句与表语从句之间的逻辑关系和衔接,译文较为松散,不符合科技文本简洁明确的特点。译文二采用综合译法,首先将原句中的第二个小句"the researchers argue"前置,译为"研究人员认为",强调动作发生的行为主体为"研究人员"。其次,将表语从句与主句结合,使用"……是……"的结构,译为"造成这一变化更可能的原因是……",译文紧凑明确,表达清晰。

(2) What makes Mike's scheme so popular is that it uses a standardized battery which fits into machines produced by different manufacturers.

译文一 让迈克的方案如此受欢迎的原因是它使用的是标准化电池可以安装在不同制造商生产的机器上。

译文二 迈克的方案之所以如此受欢迎,是因为它使用的是标准化电池可以安装在不同制造商生产的机器上。

分析 原文是一个复合长句,其中包含三个从句。第一个从句是由连接代词“what”引导的主语从句,其次是由从属连词“that”引导的表语从句,用来说明主语从句部分“迈克的方案如此受欢迎”的原因,最后是由关系代词“which”引导的定语从句,修饰先行词“a standardized battery”。对于科技文本中的此类长句,我们常采用综合译法,将原句按照汉语表达方式整合。译文一采用顺译法,首先将句首的主语从句“What makes Mike's scheme so popular is”译为“让迈克的方案如此受欢迎的原因是”,随后将表语从句和定语从句按顺序翻译。译文基本将原句中的信息翻译完整,但译文明显带有“翻译腔”,不符合汉语表达习惯。译文二采用综合译法,考虑到三个从句之间的逻辑关系和衔接方式,使用“……之所以……是因为……”句式,符合汉语表达习惯,逻辑清晰,准确流畅。

四 同位语从句的翻译

同位语从句,是指在复合句中充当同位语成分的句子,可以用来重复说明同一个称谓或事件,被解释说明的词和同位语在逻辑上存在主表关系(即被解释说明的词 = 同位语)。引导同位语从句的从属连词一般有“that,whether”;引导同位语从句的连接代词一般有“who,what,which”等;引导同位语从句的连接副词一般有“why,when,where,how”等。在翻译科技文本中的同位语从句时可以使用顺译法、转译法、综合译法等。

(1) We have come to the correct conclusion that two volumes of hydrogen and one volume of oxygen are united into one volume of steam.

译文一 我们已经得出两个体积的氢和一个体积的氧化物合成一个体积的水蒸气这样的正确结论。

译文二 我们已得出正确的结论:两个体积的氢和一个体积的氧化物合成一个体积的水蒸气。

分析 原句是由连接词“that”引导的同位语从句,进一步补充说明前面的“the correct conclusion”。译文一采用逆译法,将同位语从句前置翻译为“the correct conclusion”的修饰语“得出……样的正确结论”,意思基本完整,但句子逻辑不清,会产生歧义,不符合汉语表达习惯。译文二采用综合译法,把同位语从句译成一个独立的句子“两个体积的氢和一个体积的氧化物合成一个体积的水蒸气”,并用冒号将“正确的结论”与同位语从句分隔,表明它们之间的说明、补充关系,使得句意清晰,逻辑合理,此翻译方法适用于较长的同位语从句。

(2) The old engineer made his suggestion that the new nuclear power station should be built on a small island off the coast of Zhejiang.

译文一 那位老工程师给出了他的建议，新的核电站应建在浙江沿海的一个小岛上。

译文二 那位老工程师建议，新的核电站应建在浙江沿海的一个小岛上。

分析 原句是由连接词"that"引导的同位语从句，补充说明"suggestion"的具体内容。译文一采用顺译法和分译法，在按照原句顺序翻译的基础上，把原句分译为两个小句，这使得主句和从句的意思分散不连贯，故不符合汉语的表达习惯。译文二使用转译法，把名词"suggestion"译为动词"建议"，由此将同位语从句转译为"suggest"的宾语从句，译文前后关系明确，逻辑清晰，准确流畅。

练习题

一、英译汉

1. In physics, acceleration is the rate at which the velocity of a body changes with time.

2. If people are serious about carbon capture and storage, they will have to pay for it.

3. This was one of the observations that led a German chemist called Johann Debereiner to wonder if all chemical elements came in families.

4. The sun heats the earth, which makes it possible for plants to grow.

5. They consistently pick up highly conductive zones 2 km or more beneath the surface, for which the simplest explanation is the presence of super-salty metal rich brines.

6. Those who turn their noses up at "genetically modified" food seldom seem to consider that all crops are genetically modified. The difference between a wild plant and one that serves some human end is a lot of selective breeding-the picking and combining over the years of mutations that result in bigger seeds and tastier fruit.

7. This representation is not necessarily an accurate description of what is really going on inside the computer.

8. It has been estimated that the nuclear reactor fuel for the submarine is sufficient to last 40 circumnavigations of the globe.

9. What modern technology brings us more than happiness is the loss of human nature and deterioration of our living environment.

10. In a reheat turbine, steam flow exits from a high pressure section of the turbine and is returned to the boiler where additional superheat is added.

二、汉译英

1. 如果使用节流调节，传输效率就受到限制；然而，如果使用可调泵和电机，传输效率就很高。

2. 这让人们对铜的新来源产生了兴趣，目前大部分铜来自从大型露天矿山中挖掘出来的岩石中，然后将其碾碎加工，释放出其中所含的铜，一般约为其质量的1%。

3. 理论上,不管出口压力多大,正排量泵都能按照给定的转速产生相同的流量。

4. 人类要想利用先进技术,必须先对计算机有基本了解。

5. 为了从存于燃料中的化学能里产生有用功,首先需要通过燃烧将化学能转变为热能。

6. 在燃烧状态下,乙醇混合燃料的另一个显著差别是,当泡沫或水淹没燃烧产物时,汽油往往会先燃尽,最终留下没有可视火焰或烟雾且不易挥发的乙醇和水的溶液。

7. 尽管理论上国际单位制可以用于任何物理测量,但是大家公认有些非国际单位制仍出现在科技文本中。

8. 在某些条件下,乙醇混合燃料保留了汽油类燃料的某些特征,而在其他条件下则呈现出极性溶剂型特征。

9. 系统如何响应此类反馈,可以通过控制理论确定。

10. 较汽油机而言,柴油机燃油经济性更高,意味着柴油每单位里程所产生的二氧化碳较少。

第十二章 特殊结构的翻译

英语的句子结构多种多样，但科技英语具有专业性、精准性、客观性和名词化等特征，因而句型相对单一，但仍会使用倒装句、插入成分、强调结构等特殊结构来满足一些功能需求。

第一节 倒装句的翻译

在英语句子中，主语、谓语动词和宾语的位置一般是比较固定的。但有时为满足某些语篇功能，会调换句子的成分位置以改变语序，使句子倒装。翻译倒装句时，通常采取的翻译策略是照应译法（也称顺译法）和逆序译法（也称逆译法）。

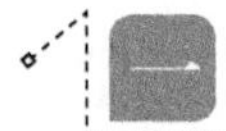

一 照应译法

照应译法就是按照原句的结构语序进行翻译的策略。

(1) On the end of the dipper arm is usually a bucket. A wide, large capacity (mud) bucket with a straight cutting edge is used for cleanup and levelling.

译文一 通常有一个挖斗安装在钟斗柄的末端，这是一个宽而且容量大的挖（泥）斗，带有平直的切口，用作清除和平整用途的部分。

译文二 钟斗柄末端通常安装一个挖斗。挖斗是一个宽而且容量大的挖（泥）斗，带有平直的切口，用作清除和平整用途。

分析 前一个小句的通常语序为"There is usually a bucket on the end of the dipper arm"，译文一就是在此基础上翻译的，不难发现语句有些生硬，该译文尝试将两个英语句子译入一个较长的汉语句子，但不符合汉语惯常表达。汉语有地点主语，如"墙上挂着一幅画"，即把地点状语放在动词前，译文二据此采取了照应译法，按照原句的语序翻译，符合汉语句法特点。

(2) A diesel engine cannot utilize all the chemical energy of the fuel, nor can any other heat engine.

译文 柴油机不能用尽燃料的化学能，别的热力机也不能。

分析 原句有由“nor”一词引导的倒装小句，表示否定前小句描述的情况，因此，按照原句的顺序翻译是符合逻辑的。“也不能”指“不能用尽燃料的化学能”这一状况，是根据语境和句法需求做了省译处理。

(3) By acceleration is meant the rate of change of velocity with time.

译文 所谓加速度，指的是速度的变化和时间之比率。

分析 “by + 名词 + is + meant”也是常见的一种倒装语序结构，意思与“acceleration means...”相同。在翻译成汉语时，语序一般不变，整个结构译为“所谓……指的是……”。

逆序译法

逆序译法是指采用与原文相反的顺序，译文采用正常的“主语 + 谓语动词 + 宾语”语序。

(1) Not so apparent is the fact that the power distribution in the sidebands is directly related to the distribution of power in the modulating wave.

译文 边带的功率分布与调制波的功率分布有直接关系，这一点并不明显。

分析 汉语有将重点强调的内容后置的习惯，英语则常用开门见山的手法叙事。汉语实际上就是结果比较突出的语言，很多表达的使用跟如何突出结果有关。也就是说，汉语是“先事实，后评价”，英语则是“先评价，后事实”，本句译文旨在突出这一特点。

(2) The studies have also highlighted that as the saturation water content of soils varies dramatically, so will the absorption of oxygen in the soils.

译文 研究还表明，土壤的饱和含水量差异巨大，其含氧量差异亦然。

分析 英语原文如果以“so, hence, thus, there, then, often”等副词开头时，经常会用到倒装语序，汉译时，要将语序还原成自然语序。本句中，还需要根据语境将“so”的具体意思还原。

(3) 历史上炼金术士们寻求长生不老药的企图从未实现过。

译文一 The alchemists in history never realized their attempts to find a whimsical substance called “the elixir of life”.

译文二 Never did the alchemists in history realize their attempts to find a whimsical substance called “the elixir of life”.

分析 译文二为了强调从古至今从未有人找到长不老药这一事实，将原文倒装，起到了突出重点的作用。同时，增译“whimsical”以描述“长生不老药”这一物质所具有的虚幻特征。但值得注意的是，个别句子根据上下文语境用顺译法和逆译法均可,因此译文一在一定语境下也是可接受的。

(4)工地中央有一座塔吊,正在把预制构件往一座高楼的构架上起吊。

译文一 At the center of the construction site stands a tower crane which is lifting the prefabricated elements and parts onto the frame of a high-rise building.

译文二 A tower crane at the center of the construction site is lifting the prefabricated elements and parts onto the frame of a high-rise building.

分析 译文一按照原文的叙述顺序处理,运用倒装结构先翻译了地点状语"At the center of…",而后由"which"引导的定语从句描述了塔吊的具体工作内容,强调了地点。译文二则遵照常规英语叙事结构"主语+谓语动词+宾语","at the center of the construction"是主语"a tower"的地点状语。联系上下文,可以根据语境从两个译文中选择其一。

(5)这种放射性元素无论如何不能暴露在空气中。

译文一 This radioactive element should not be exposed to the open air under no circumstance.

译文二 On no account should this radioactive element be exposed to the open air.

分析 译文一译出了原句的基本含义且语法正确。译文二使用倒装结构突出了原文,强调了化学元素暴露的潜在危害性。另外值得注意的是,原文"暴露在空气中"隐含了被动意义,因此英译时应使用被动语态。

第二节 插入语翻译

插入语本身是一个结构独立、意义完整的成分,将其插入一个语法完整句子中的词、短语或句子之间。插入语可以置于句首、句中或句尾,如:

Unlike armor, which is intended to limit damage to a vehicle if it is hit, APSS are there to stop missiles striking in the first place.

译文 装甲旨在减少车辆在被击中时受到的损害,与装甲不同的是,APSS是为了在一开始就阻止导弹袭击。

分析 由"which"引导的非限制性定语从句表示举例、附加解释或说明,在此句中是对"armor"的进一步补充和说明,翻译时可以采取直译法,符合汉语表达,行文也流畅自然。

针对插入成分的翻译方法,刘宓庆指出,宜采用包孕、拆离等方法按照自然语序翻译,以免由于间隔过长,切断主干脉络,使读者在阅读中产生阻断感。

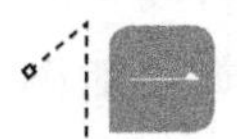

包孕译法

所谓"包孕",就是将英语译成汉语时把英语中的后置修饰成分放在中心词之前,使修饰成分在汉语句中形成前置包孕。汉语更依赖词序来表达思想,句内成分的排列顺序往往和我们观察、感知世界的顺序相对应,因此,采用包孕译法能使译文更加符合汉语表达方式或习惯。

(1) Between these two tiny particles, the proton and the electron, there is a powerful attraction—the attraction that is always present between negative and positive electric charges.

译文 质子与电子之间有一种强大的引力——正负电荷间总是存在着这种引力。

分析 原文使用逗号插入"two tiny particles"的同位语"the proton and the electron"对其做进一步解释,在汉译时,译文将其放在中心词之前以修饰该中心词,使得译文的表述更加完整清楚。

(2) Although you cannot quite see it from the side walk, the Aria's main entrance is anchored by not one but two fountains, both from the world-renowned water design firm WET.

译文一 尽管从人行道上看不见,但阿瑞尔酒店的正门处有两个喷泉,都是由世界著名喷泉设计公司 WET 设计的。

译文二 尽管从人行道上看不见,但阿瑞尔酒店的正门处有两个由世界著名喷泉设计公司 WET 设计的喷泉。

分析 原句是由"Although"引导的让步状语从句,"both from the world-renowned water design firm WET"用作同位语解释说明"two fountains"。译文二采用包孕译法,将同位语部分译成前置定语,使译文结构紧凑,表达流畅。而译文一将原文中的定语从句单独翻译成一个小句,这样不仅不符合汉语的表达习惯,更有碍读者获取文本信息。

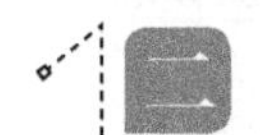

插入译法

插入译法就是利用破折号、括号或前后逗号将句子成分插入译文中。这样一来,既保持句子原有的信息焦点,又确使句子准确传达全部信息。插入成分的主要文体功能有简化语言结构、突出篇章意义和避免多重句子重心。

(1) Hence, it would be difficult to explain why, if perception is as multisensory as we are now being led to believe it is, that the Gestalt principles should affect perceptual organization in certain senses but not others.

译文一 所以说,很难解释,倘若确实如上所述——知觉是多感官的运用,那么为什么格式塔原则单单作用于某些感觉官能的知觉组织,而对其他感觉官能倒不起作用。

译文二 所以说,倘若确实如上所述——知觉是多感官的运用,那么为什么格式塔原则单单作用于某些感觉官能的知觉组织,而对其他感觉官能不起作用。对此人们很难做出解释的。

分析 原句有由"if"引导的条件状语从句,汉语侧重结构描述,如果将这一从句翻译为"如果知觉就像我们现在认为的那样",而在上文未做具体情况的描述,这样就会造成译文含义模糊。补充破折号后的内容——"知觉是多感官的运用"后,使得译文传达的内容更加完整。另外,对比两个译文,"很难解释"作为对前述内容的评价,译文二将其放在句末是符合汉语"先事实后评价"的表达习惯,比译文一更具逻辑性。

(2)热固塑料,一旦成形,就不再改变其形状了。

译文 A thermosetting material, once it is formed, will no longer change its shape.

分析 "一旦成型"在原句中是表示条件的插入成分,汉语结构比较松散,这样的插入成分较为常见。译文遵照原文的语序插入了一个条件句,不影响句意的表达。另外,译文中的"once it is formed"还可以省略成"once formed"。

三 拆离译法

英语中的插入成分会增加句子的长度和难度。英译汉时,顺译长句中的某些成分有时会造成译文不畅。拆离就是将长句中某些成分(词、词组或句子)从句子主干中拆开,另行处理,以利于句子的总体安排。可以将这些子句置于句首或句末,以实现"化长为短,化整为零"。

(1) In some places the ground may be so strong that roadways will stand almost indefinitely without artificial support, as will be seen from Table 16, which shows that about 9% of underground roadways are underlined.

译文 在有些地方,地层很坚固,虽然未用支护,但巷道几乎无限期地保持着。如表16所示,大约9%是未支护的裸体巷道。

分析 本句为多重复句,主句是"so…that…"引导的状语从句,从句内又包含了一个"as"引导的定语从句插入到句子中来指代前面整个主句,其后引导的从句是实例句,与主句联系不太紧密。经此分析后,可以考虑在保留原文信息完整的情况下,将整个长句拆译为两个汉语独立句,即由"as"引导的那部分插入语可译入另外的小句中,这样处理有助于清晰传达原文的含义。

(2) One reason for its infrequent use, except where it is the only practical solution, is that the almost constant use of thrusters to maintain position requires large amounts of fuel, adding significantly to operating costs.

译文一 不常用的原因之一,除必须使用该系统解决实际问题之外,是持续使用推进器保持方位需要大量燃油,这极大地增加了操作成本。

译文二 除必须使用该系统解决实际问题之外,不常用的原因之一是持续使用推进器保持方位需要大量燃油,这会大大增加操作成本。

分析 句子主干为"One reason for its infrequent use is that…",但要注意句子之间的逻辑关系,以及"except"后接的小句和现在分词"adding"作为结果状语的成分。对比两个译文,后者将"except"后的小句拆离翻译在句首,剩余部分则按照原文翻译,行文符合汉语表达习惯;而前者翻译顺序完全遵照原文,译文不流畅自然。

四 重组译法

科技英语使用长句较多,有时需要综合考虑对长句的处理方法,按时间顺序,或按逻辑顺序,做综合处理,把英语原文翻译成通顺忠实、主次分明的汉语句子。重组是指将长句结构完全理清后,按汉语叙事习惯重新组合句子,这摆脱了原文的语序和形式的约束,使译文流畅、自然。

(1) The heads of all the set-screws and bolts should, if possible, be made the same size.

译文 如可能,所有螺钉和螺栓的头部,都应按统一尺寸制作。

分析 原句的主语和谓语之间有插入语并由逗号和破折号隔开,破折号后的内容表示解释说明,可以与主语"The heads of all the set"糅合。汉译时将"if possible"处理为表示条件的状语从句,使之位于句首,重组句序后更加符合汉语的表达习惯。

(2) 这种燃料电池车与汽油车和柴油车不同,不排放二氧化碳和氮氧化合物等有害物质,排出的仅仅是水。它不仅环保,而且能源利用率高,因此人们认为它是21世纪的理想交通工具。

译文 Unlike its petrol or diesel counterparts, this fuel cell vehicle emits water instead of harmful substances like carbon dioxide and oxynitride. For its eco-friendliness and fuel efficiency, the vehicle is considered an ideal means of transport for the 21st century.

分析 译文用"counterparts"指出了新能源车辆与传统车辆有不同之处,"instead of…"这一短语结构也让它们形成了对比,译文注重了这一逻辑关系,同时将形容词"环保"转译为名词,句子更为简洁明了。

第三节　强调结构的翻译

在英语中,有时为了凸显某些句子成分,增强语气,会采用强调结构。“It is/was…”是英语中最常见的强调句型,强调部分常由“that/who/which”引导,而后再加上其他组成部分。

一　强调主语

当强调的主语是人时,常用“who”或“Whom”引导,如:“It was John whom/who I saw yesterday”, 主语是物时,常用“that”或“which”表示强调。

(1) We conclude that it is the ratio of the average signal power to the average noise power that is important, and not the magnitudes of signal and noise themselves.

译文　我们的结论是,平均信号功率与平均噪声功率之比相当重要,而信号与噪声本身幅度的大小并不重要。

分析　原句所强调部分的主干结构是“it is… that is important…”,在此从句中所强调的是主语“the ratio of the average signal power to the average noise power”。译文按照原句所呈现的语序翻译,将强调部分单独译成小句,并且增译了“并不重要”,从而更完整地传达全意。

(2) 正是科学家们经过无数次实验才克服了这一困难。

译文　It was scientists who overcame this difficulty through innumerable tests.

分析　原句强调的是主语“科学家们”。汉语可以通过调整语序或者增添如“正是”和“确实”一类的词语来起到强调的作用。而在翻译成英语时,就需要借助一些固定的结构来凸显成分。

(3) 正是拉伸应力使构件失去其承载功能。

译文　It is the tensile stress that causes the component to fail in its load-carrying function.

分析　原句可以理解为“使构件失去承载作用的就是拉伸应力”。译文中的“正是”二字强调了主语“拉伸应力”(the tensile),准确传达出原文信息,句子结构,流畅自然。

二　强调宾语

(1) There is a growing interest in the idea of using photoelectric cells to run gadgets as well

as power grids, and doing so even when those gadgets are inside buildings. It is the photoelectric cells that G24 Innovations, a firm based in Wales, have come up with.

译文 人们对于利用光电池来驱动用电装置和电网越来越感兴趣了,这些光电池即使在室内也能够正常工作。这正是某总部位于威尔士的公司——G24 Innovations,开发出来的光电池。

分析 原文第一句介绍了光电池这一新的发明,而后使用强调结构突显这是由“G24 Innovations”发明的光电池,宾语从而得以强调。译文将原文所强调的逻辑宾语——“光电池”(the photoelectric cells)单独译入最后一个小句中,旨在凸显这一宾语。

(2) Such capture is the first part of a three-stage process known as carbon capture and storage (CCS) that many people hope will help deal with the problem of man-made climate change.

译文 很多人希望利用这种称为碳捕捉与储藏技术(CCS)的三段式方法来应对人为的气候变化问题,而这种捕捉是其中的第一阶段。

分析 “Such (a)...is...”这一结构也可以用来强调,原文强调的是表语——“the first part of a three-stage process”(三段式方法的第一阶段)。译文将其从句子中拆离出来翻译,在末尾单独成句,旨在更能达意。

(3) 在混凝土中起化学作用的是水泥而不是集料。

译文 It is the cement but not the aggregate that reacts in concrete.

分析 “……起化学作用的是……而不是……”这一谓语成分后强调的是宾语——“水泥”(the cement)。译文使用“It is...that...”这一典型强调结构确切地传达出了原文的含义。

三 强调状语

(1) It was not until Madame Curie discovered radium that people had some idea of what radioactivity was.

译文 直至居里夫人发现镭,人们才对放射现象有所了解。

分析 “It is/was not until...that...”是较为常见的强调时间状语的句型,强调的内容可还原成一般句型——“People had some idea of what radioactivity was after Madame Curie discovered radium.”。译文遵照原文行文结构,将这一强调时间的结构汉译为“直至……才……”,是忠实于原文的翻译策略。

(2) It is by the use of steam-jackets that a reduction in condensation is achieved.

译文 正是使用蒸汽套才起到削弱冷凝过程的作用。

分析 原句强调的是方式状语“by the use of steam-jackets”,汉语译文通过词汇“正是”来表达这种强调的语气。

(3)无论如何,机器的输出功率绝不可能大于输入功率。

译文 Under no circumstance can more power be got out of the machine than is put into it.

分析 “无论如何”是条件状语,译文遵循了原句的结构顺序,先翻译了这一状语,“under no circumstance”这一完全否定的用法有力地强调了句子的状语。

四 强调其他成分

除了主语、宾语和状语之外,整个句子或者句内的从句也可以得到强调和凸显。

(1)Very obvious is the ever-growing influence on mankind by the adoption of radio-broadcasting both sound and television.

译文 很明显,无线电及电视广播的应用日益影响着人类。

分析 原文强调的是整个句子所描述的内容。译文采用顺译法还原了这一结构,将起到强调作用的“很明显”置于句首,使原句整个句子得到强调,译文读者也能获取到这一信息。

(2)他们确实试用各种方法来改进这台电动机的性能。

译文 They did try every means to improve the performance of the motor.

分析 汉语原句是对整个句子所描述情况的肯定强调,译文通过增添英语助动词“did”——相当于汉语中的副词,较为完整清晰地传达了这一意思。

(3)But for some aficionados it is only now that artificial intelligence (AI) can truly say it has joined the game-playing club—for it has proved it can routinely beat humans at Diplomacy.

译文 但就一些狂热爱好者而言,AI 现在才算是真正加入了游戏俱乐部——因为已在《强权外交》游戏中它证明自己通常能击败人类。

分析 “It is now that...”是强调时间状语的典型句型,译文将“that”后的逻辑主语“AI”翻译在前,随后是时间,这与汉语主语先行的原则相符,便于译文读者阅读。

练习题

一、英译汉

1. There is a nucleus at the center of an atom. Around this nucleus revolve electrons.

2. No matter what the nature of energy is, the total energy remains unaltered in quantity but may change in form.

3. Quartz is common on Earth but exceedingly rare on Mars, indicating that granite, from

which it forms, is scarce. Nor is there evidence for metamorphic minerals such as slate or marble, produced when volcanic or sedimentary rocks are subjected to high pressure or temperature.

4. However, the position of the eyes so restricts the field of vision in baleen whales that they probably do not have stereoscopic vision.

5. Passive smoking, the breathing in of the side-stream smoke from the burning of tobacco between puffs or of the smoke exhaled by a smoker, also causes a serious health risk.

6. It is not yet known whether robots will one day have vision as good as human vision.

7. Today it is finished manufactured products that dominate the flow of trade, and thanks to technological advances such as lightweight components, manufactured goods themselves have tended to become lighter and less bulky.

8. It's very important to take care of your health, but there is no need to fear too much of diseases.

9. It turns out that the global average temperature is quite stable over long periods of time, and small changes in that temperature correspond to enormous changes in the environment.

10. It is desirable that the biomaterial elastic modulus is similar to bone.

二、汉译英

1. 黑色森林里有很多湖,这些湖里游动着各种各样的鱼。

2. 钻石虽然很硬,但在强大的压力下也会被压得粉碎。

3. 把物体举得越高,它所获得的重力势能也就越大。

4. Wohler 于 1827 年在海德堡大学里发现了一种新的金属元素。

5. 有必要找到一种能吸收中子的材料,实现停止连锁反应,或使其保持在释放出的能量不致使系统熔化的水平上。

6. 正是由于这些原因,这种"高速公路"必须建成共同合作的特别工作组,而不是过去使用的更为传统的服务供应商/用户模式。

7. 人们普遍认为陆地动物是由海洋动物进化而来的。

8. 应当强调指出,风能也是自然界中可以利用的清洁能源。

9. 由此可以得出结论,对于电话工程师而言,波特率非常重要,因为它决定了所有电信信道的类型。

10. 人们普遍认为,汽车和其他机器已经成为我们社会必不可少的一部分。

第四部分

科技语篇译例分析

第十三章
机电工程类

【英语原文】

Circuit Diagram for the Electrical Field Detector (excerpt)

The simple proximity detector whose circuit diagram is shown in Fig. 4 exhibits some unique features[1]. Unlike[2] other techniques such as infra-red, ultra-sonic, light-activated, microwave doppler[3], etc.[4], it does not use a sensing device other than a short piece of wire that acts as the pickup[5]. It is omni-directional[6] and requires no setting up. It will detect movement up to a distance of ten feet, subject to environmental conditions[7].

Electric fields exist almost everywhere in various patterns and strengths, depending to a great extent on conducting objects, including people and animals, within the immediate area[8]. If left undisturbed, the field changes very slowly over a long period of time but if, say, a human body[9] moves through this static field, it causes turbulence, resulting in a change[10] in the geometry and strength of the field.

Sensitivity

The degree of sensitivity of this instrument is subject to the environment in which it is located. It is especially excited by synthetic materials found in furnishings and clothing, such as vinyl and polyester[11]. The device therefore adjusts to the surroundings within its operating range and these conditions become its static reference against which it compares any fairly fast changes that may occur[12].

If, now, a person comes within its sensing range and stands still[13], the device will signal the change and then return to its quiescent state after adjusting to the new conditions, that is, the modified field strength and pattern[14]. Because the signal is only fleeting, it is necessary to capture it through a pulse stretcher[15] to make it of practical use.

Circuit Details

In this design an antenna, comprising a short length of wire, insulated or bare[16], is connected in an upright position to the gate (G) of a field effect transistor, TR1, which has very high input impedance and a low output impedance[17]. TR1 therefore amplifies the field potential detected by the antenna[18]. The next stage is a 50Hz a. c. filter which screens out most of the a. c. field from mains wiring[19].

The output from this filter is split into two paths[20]. One is connected to the inverting input of voltage comparator IC1; the other is connected to the non-inverting input via a low-pass filter[21]. This filter, comprising R5 and C3, serves as a reference voltage against which the signal voltage arriving at the inverting input is compared[22]. The filter accepts only very slow voltage changes but blocks fairly rapid changes resulting from a moving conducting body within its sensing area[23].

The amplified signal at the output of the comparator switches on transistor TR2 via coupling capacitor C4 and buffer resistor R6[24]. This in turn triggers a simple R/C timing circuit (C5 and R8) to operate a relay or other device via Darlington transistor TR3[25]. The stated values of C5 and R8 provide a delay of approximately ten seconds, but one or both values can be increased for longer periods[26].

No setting up is involved[27] but there is plenty of room for experimentation, particularly in regard to the antenna[28]. Generally, one having a length of six to eight inches will give satisfactory results[29]. To test the instrument, attach a voltmeter to the output, preferably with a load, and switch on. Stand well back and remain still[30].

The meter should immediately indicate a voltage close to the supply rail, which will slowly decay to zero after about one minute[31]. The instrument is now primed to detect any movement within its radius of operation[32].

【文本分析】

原文节选自《电场探测器电路图》,文中技术类词语较多,语体正式(formal),语域(register)中等,句式较为简单。该文本属于纯信息类文本,理想的翻译风格为"中立、客观",翻译方法为等效翻译,最大翻译单位为句子,最小翻译单位为搭配。该文本翻译的重难点为技术词语的翻译和个别复杂句式的处理。

【原译】

电场探测器电路图(节选)

其电路图如图4所示的简单接近检测器表现出一些独特的特征[1]。与其他技术不同[2],如红外线、超声波、光激活和微波多普勒[3][4],除了充当拾音器的一小段电线外,它不使用传感设备[5]。它是全方位的[6],无须设置。根据环境条件,它可以检测最远十英尺距离的运动[7]。

电场以不同模式和强度几乎无处不在,这在很大程度上取决于附近区域内的导电物体,包括人和动物[8]。如果不受干扰,场在很长一段时间内变化非常缓慢,但如果人体在这个静态场中移动[9],就会引起湍流,导致电场的几何形状和强度的变化[10]。

灵敏度

本仪器的灵敏度取决于其所在的环境。它对家具和服装中的合成材料特别感兴趣,例如乙烯基和聚酯[11]。因此,该设备会根据其工作范围内的环境进行调整,并且这些条件成为其静态参考,它将可能发生的任何相当快的变化与之进行比较[12]。

现在,如果一个人进入其感应范围内并静止不动[13],该设备将发出变化信号,然后在适应新条件(即修改后的场强和模式)后返回其静止状态[14]。由于信号转瞬即逝,因此需要通过脉冲展宽器[15]将其捕获以使其具有实际用途。

电路详细信息

在这种设计中,包括绝缘或裸露的短长度电线的天线[16]以直立位置连接到场效应晶体管 TR1 的栅极(g),该晶体管具有非常高的输入阻抗和低的输出阻抗[17]。TR1 因此放大由天线检测的场电势[18]。下一阶段是 50Hz 交流过滤器,该过滤器将大部分交流电场从总线中屏蔽出来[19]。

该过滤器的输出分为两条路径[20]。一个连接到电压比较器 IC1 的反相输入;另一个通过低通滤波器连接到非反相输入[21]。该过滤器包括 R5 和 C3,用作参考电压,与到达反相输入端的信号电压进行比较[22]。该过滤器只接受非常缓慢的电压变化,但屏蔽其感应区域内移动导电体导致的相当快速的变化[23]。

比较器输出的放大信号通过耦合电容器 C4 和缓冲电阻器 R6[24]接通晶体管 TR2。这进而触发一个简单的 R/C 定时电路(C5 和 R8),通过达林顿晶体管 TR3 操作继电器或其他设备[25]。C5 和 R8 的规定值提供了大约 10 秒的延迟,但一个或两个值可以在更长的时间内增加[26]。

不需要设置[27],但有大量试验的空间,尤其是关于天线的试验[28]。通常,长度为 6 ~ 8 英寸的天线会给出令人满意的结果[29]。要测试仪器,最好在输出端连接一个电压表,最好带有负载,并打开[30]。站好,保持静止[31]。

仪表应立即指示电源轨附近的电压,大约一分钟后,电压将缓慢降至零[32]。仪器现在已准备就绪,可以检测其操作半径内的任何移动[33]。

【参考译文】

电场探测器电路图(节选)

如图 4 所示,此简易接近式探测器的电路图具有一些特性[1]。该探测器不像[2]红外、超声波、光敏、微波、多普勒[3]等[4]技术使用传感装置,而是只用一根短线作为传感器[5]。该探测器可全方位感应[6]、无须设置,最远可探测 10 英尺内的物体移动,具体视环境情况而定[7]。

电场几乎无处不在，且形式不定、强度不同[8]，这很大程度上取决于周边的传导体，包括人类和动物。如果不受干扰，电场在长时间内变化很慢，但若有干扰体（如人体）[9]穿过静电场，就会引起湍流，从而改变[10]电场的几何结构和强度。

敏感度

该仪器的敏感程度受其所在环境的影响，特别易受室内陈设品和衣物中的乙烯基、聚酯等合成材料的影响[11]。因此，该仪器能在其运作范围内调整，以适应周围环境。这些条件为与之相比可能发生的较快变化提供静态参照[12]。

如果现在有人来到仪器感应范围内，站立不动[13]，仪器将发出场变信号。仪器适应新电场环境（即改变后的场强和场形）后，信号将恢复静止状态[14]。由于该信号转瞬即逝，因此有必要用脉冲展宽器捕捉信号[15]，使其具备实际效用。

电路详情

本设计中，一根天线（由一根短绝缘线或短裸线构成[16]）垂直连接在一根输入阻抗极高、输出阻抗较低的场效应晶体管（TR1）的栅极（g）上[17]，因此可放大天线接收到的场电位[18]。接着，一个50赫兹的交流电滤波器将干线产生的多数交流电场屏蔽[19]。

交流电滤波器的输出分为两路[20]。一路接到电压比较器IC1的反相输入端，另一路经过一个低通滤波器接到电压比较器的同相输入端[21]。低通滤波器由R5和C3组成，用作参考电压，以比较到达反相输入端的信号电压[22]。低通滤波器仅允许低频电压通过，阻止其感应范围内移动传导体引起的高频电压通过[23]。

比较器输出端放大的信号通过耦合电容器C4和缓冲电阻器R6接通晶体管TR2[24]，继而触发一个简单的R/C定时电路（C5和R8），通过达林顿晶体管TR3操作继电器或其他设备[25]。上述C5和R8的值提供近十秒的延时，但可以增加一个或两个值以获得更长的延时时间[26]。

本探测器不需要设置[27]，但还有调试的空间，特别是调试天线[28]。一般来说，天线长6～8英寸时，效果较佳[29]。为测试仪器的效果，可在输出端连接一个电压表（最好有负荷），然后打开[30]。退远站好，保持不动[31]。

测试过程中，电压表应立即指示导轨附近有电压，电压约一分钟后慢慢衰减到“0”[32]。如果是这样，探测器就已经准备就绪，可探测其操作半径内的任何移动[33]。

【译文详解】

[1] 这句话中有一个定语从句，原译根据英语语序直接翻译为“其电路图如图4所示的简单接近检测器表现出一些独特的特征”。此译文明显受原文束缚，佶屈聱牙。一般情况下，将较短的定语从句译为汉语时，修饰成分放在中心词前；较长的大多需要拆分，单独成句，如译为“此简易接近式探测器的电路图如图4所示”。本句从句虽短，但考虑到标题为“图4”，可译为“如图4所示，此简易接近式探测器的电路图具有一些特性”。

[2] 当“unlike”“different from”“similar to”等结构置于句首时，一旦后面衔接的内容较多，汉语流畅度就会受到影响。通常的做法是，将后面的主语挪至这些结构前，译出完整的句子。参考译文中这句话的处理比较灵活。从“unlike”和“does not use”可以推知，前面几项技术是使用传感装置的。因此，译为“不像……那样使用传感装置，而是……”。

[3] 通过网络查询工具得知，探测仪或探测器分为红外探测仪(器)、超声波探测器、光敏探测器、微波探测器、多普勒探测器等。因此，可以断定原文“microwave doppler”中间应该有个逗号。另外，“仪器”和“技术”二词有较大差异。例如，用于探测器时可以说“红外线探测器”，也可以说“红外探测器”，而用于技术语境时，一般说“红外技术”而不说“红外线技术”，翻译时要格外留意。

[4] “such as”和“and so on”“etc.”不能连用，原文用法有误。值得注意的是，实际翻译工作中遇到原文有语法错误、逻辑谬误等问题的情况实属常事。汉语用一句话表达“总称+列举”时，通常是“先列举、再归纳”，如“当前国内的多样性算力发展面临着标准体系不成熟、评测基准不完善、生态发展薄弱等多方面挑战”。因此，这句话尽量不要按照原句式翻译为“其他技术如红外、超声波、光敏、微波、多普勒等”，而应译为“红外、超声波、光敏、微波、多普勒等技术”。

[5] 根据https://dictionary.cambridge.org/dictionary/english/other-than的解释，“other than”意为“except”；“not...other than”意为“除……外，没有”，表强调，因此译文增译了“而是只”。“sensing device”应译为“传感器”。拾音器，又称监听头。监听拾音器是一种用来采集现场环境声音再传送到后端设备的仪器，而本文语境中不涉及采集声音。

[6] 英语忌讳重复，因此多使用代词去替代前面提到的名词。汉译时，需要将其中一部分代词还原为具体的名词。因此，此处的“it”译为“该探测器”更为妥帖。“omni-directional”是形容词，若直译为“该探测器是全方位的”，会让人不知所云。此处采用词性转换法(conversion)，将静态的英文表达转换为动态的中文表达。

[7] 原文最后两句内容密切相关，汉语可以用逗号连接。当然，若处理为两个句子，不妨将subject to这部分前置，译为“该探测器可全方位感应、无须设置。视环境情况，最远可探测10英尺内的物体移动。”

[8] (1)“patterns”有“形式”“模式”“模型”等意思。可在语料库中分别检索“电场模型”“电场模式”“电场形式”等并结合下文的“50Hz”“AC”(交流电)等加以验证。(2)原译将这句话前半部分处理为“电场以不同模式和强度几乎无处不在”，这种表达主要是受欧化中文的影响，不符合中文表达习惯。“and”并列的成分如果可以用汉语四字格表达，通常不用译出“和”；加之“various”修饰两个名词，因此译为“形式不定、强度不同”更为简洁、地道。当然，译为“电场几乎无处不在，其形式和强度各不相同”也很通顺。(3)“conducting objects”原译为“导电物体”，但“导电”一般需要肉眼可见的接触。根据

英汉词典的解释,“conduct”本意为“传导”这一物理术语,只是在某些具体语境中译为“导电”。综上,此处译为“传导”更为准确。

[9] 原文的“if, say, a human body”比较口语化,直译为“但如果,比如说,人体”不够正式。此处的难点在于如果“什么”穿过电场。从上下文推断,是“干扰体”,故需要增译“干扰体”一词。而后将“如:人体”置于括号内,更为正式、清晰。另外,“field”是指“electric field”,汉译时最好也将其增译出来。

[10] 原译将“resulting in a change”译为“导致电场的几何形状和强度的变化。”,稍显拗口。可增译“发生”一词,译为“导致电场的几何形状和强度发生变化”,更符合汉语动态性的特点。需要注意的是,“result in a change of”或“make a change of”本身的意思就是“change”,英语有较多弱势动词+名词的结构,这在英语内圈国家也饱受诟病。而汉语中“进行”“作出”等万能动词+名词的结构也多为欧化而来,而纯正的汉语往往惜字如金,因此参考译文直译为“改变”。想要深入研究,可参阅琼·平卡姆、姜桂华所著的《中式英语之鉴》及余光中先生所著《中文的常态与变态》或《余光中谈翻译》。

[11] (1)参考译文将本段第一、二句话合译为一句话。值得注意的是,很多人将“especially”译为“尤其(是)”,而地道汉语更倾向于使用“特别是”;若“especially”后面内容较多,还会加上“尤其如此”“更是如此”“尤其明显”等表达。综上,第二句话还可以译为“特别是室内陈设品和衣物中的乙烯基、聚酯等合成材料对其造成的影响,尤(更)为明显。”(2)“such as”的处理方法请参考本章[4]。

[12] 参考译文将“become its static reference”译为“提供静态参照”更符合中文的表达习惯。如果后面接有宾语,汉语通常使用“为……提供静态参照”的句式。这句话还可以分译为“这些条件为仪器提供静态参考,与可能发生的任何较快的变化相比较”。

[13] 参考译文中的“有”为增译而来,增译后更符合汉语表达习惯;“and”可以省译,不过译出来也无伤大雅。

[14] (1)原译将“new conditions”译为“新条件”,不够具体;参考译文译为“新电场环境”,更清晰明了。(2)“field strength and pattern”中“field”修饰两个名词,译为汉语时需要出现两次,即场强和场形;绝对不能将其译为“场强和模式”或“场强和形式”。(3)原文看似为“the device will then return to its quiescent state”,但实则仪器(即探测器)本身并不会动,会发生变化且恢复至静止状态的是电场信号,即原文中的“signal”。因此,参考译文基于技术原理增译了“信号”一词。

[15] “pulse stretcher”的翻译比较麻烦。可能的译法有“脉冲展宽器”“脉冲扩展器”“脉冲伸张管”“脉冲伸张器”等。经查证,脉冲扩展器,即“脉冲稳定器”,英文为“pulse sustainer”,因此排除。在百度和谷歌中查看词条数量,“脉冲展宽器”的词条明显多于其他词条,并且还有与该词条相关的研究性论文。综上,可以确定“pulse stretcher”的

译文为“脉冲展宽器”。

[16] 原译十分迂回。“a short length of wire, insulated or bare”意为“不管是绝缘的短线还是裸露的短线”,但这个表达仍然比较冗长且不符合行业说法。用汉语搜索引擎一一验证可知,较为贴切的说法是“短绝缘线”或“短裸线”。注意“short”限定后面两个形容词,因此翻译时需要增译一个“短”字。

[17] (1)“is connected in”译为汉语时宜省略“被”字;“upright”在字典或机译平台中的结果可能是“竖直”或“垂直”,可输入“竖直连接”或“垂直连接”(加上引号强制搜索)加以验证;“field effect transistor”在网络词典中有“场效应晶体管”“场效晶体管”“场效应管”等翻译,经查询,“场效应晶体管”简称“场效应管”,这两者最为常用;“gate”机译的结果是“栅极”,再在语料库中检索“场效应晶体管 G”加以验证,十分容易确认。(2)“which”引导的非限制性定语从句更多译为状语从句,或者单独成句用以补充说明。此句的“TR1”是“field effect transistor”的名字,因此可以理解为限定修饰。值得注意的是,如果此处没有“TR1”字眼且“which”引导的是非限制性定语从句,还应考虑“which”是否是一种“垂直连接”的结果,再进一步验证连接方式不同与输入阻抗和输出阻抗的高低是否真的有关。

[18] “TR1 therefore amplifies”是基于前面整句话而来的,这再次证明,上一句的“which”应理解为限定成分。此句和上一句有因果关系,中文可用逗号连接。“field potential”可译为“场电势”或“场电位”,经查证,“电势”和“电位”基本上是同一个概念,在电路中多用“电位”。

[19] (1)“The next stage”的翻译较为灵活。不宜译为“下一阶段”,其原因是:一个阶段通常有一连串的行为,而此前的表述中未体现相关含义。也不宜译为“下一步是……滤波器,它将……”,这种表达前面半句缺少动词,是典型的死译。(2)“mains”在这里作形容词,意为“source of...”;这一用法较少使用,位于词典释义中最后一条。“mains wiring”的意思为“总输电线”或“干线”。(3)“filter”可能译作“过滤器”,也可能译为“滤波器”,要根据后文判断。

[20] 本句中的“the filter”指的是上文中的“交流电滤波器”,为了跟下文的“低通滤波器”区分开来,需要增译“交流电”;“is split into”省略“被”字,译为“分为”;“two paths”原译为“两个路径”,但在此处不符合语境。

[21] “low-pass filter”很容易译为“低通滤波器”。“inverting input”直译为“反向输入”,但在本句中明显不通;此处应译为“反向输入端”。同理,“non-inverting input”应译为“同向输入端”。

[22] (1)“This filter”指的是“低通滤波器”,需要增译“低通”二字;“comprising”相当于“which comprises”。这部分用两个逗号隔开,用以补充说明,汉译时直接顺译即可;译为“这个由R5 和 C3 组成的低通滤波器”反而拗口。值得注意的是,本文中有很多“the /this + n”

结构,如果没有歧义,可省略“该”或“本”字。(2)“reference voltage”可能译为“参考电压”或“参照电压”,在汉语搜索引擎中搜索很容易得知,业界一般用“参考电压”。(3)“is compared”译为主动式。

[23] (1)“accepts”一般为“接受”之意,但“接受电压变化”在学科语境中显得不够专业。通过搜索“低通滤波器”和“低通滤波器的工作原理”得知,“accepts”译为“允许”更为恰当,当然“允许”后面一般要加动词,故增译“通过”一词。但是,“允许慢的电压变化通过”仍然令人不知所云,而查询相关背景知识得知“高频电压变化快,低频电压变化慢”。综上,“The filter accepts only very slow voltage changes”译为“低通滤波器仅允许低频电压通过”。(2)“blocks voltage change”译为“阻止”而不是“屏蔽”电压变化,这也需要查阅背景知识。“导致”一词偏贬义,在科技文体中译为“引起”更合适。

[24] “output”要根据语境增译“端”,使其更为具体、清楚。“输出端的放大信号”稍显别扭,改译为“(在)输出端(被)放大的信号”。“switch on”有“接通”“开启”“打开”等之意,可分别将三个动词搭配“晶体管”加双引号搜索,选用搭配量大的译法。

[25] “in turn”在这里的译法比较灵活,译者可自行斟酌。“timing circuit”可能译为“定时电路”“计时电路”“时序电路”等。

[26] (1)注意被动语态汉译时的顺序,此处采用逆序法。同时后半句应理解为“sth. can be increased, for longer periods”或“for longer periods, sth. can be increased”,而不是“increase by/to”。(2)“for”表示目的,汉译时多将词性转换为动词。(3)原译的“时间”不够具体,参考译文根据上文增译了“延时”一词。

[27] “no…is involved”一般译为“不涉及”,但“不涉及设置”在缺乏语境情况下难以理解,此处根据需要增译“本探测器”。另外,“探测器不涉及设置”搭配不当,因此需要进一步改为“本探测器不需要设置”。

[28] (1)此处首先应该注意“试验”和“实验”的区别:“试验”侧重“试用”,而“实验”侧重“认识自然现象、自然性质、自然规律”。在该语境中,“试验”比“实验”更为恰当。但“有试验的空间”这一表达不太符合探测器行业说法,因此参考译文改为与其意思相近但更专业、更准确的“调试”。(2)“particularly in regard to the antenna”原译为“尤其是关于天线的试验”,相对拙劣,欧化痕迹明显,参考译文的“调试天线”更加简单明了。

[29] (1)“one having a length of six to eight inches will give satisfactory results”原译为“长度为6~8英寸的天线会给出令人满意的结果”,表达较晦涩。此处参考译文将“天线”这一旧信息提前,作主语,同时增译“时”,将原句的主语分译为条件分句“天线为6~8英寸时”。(2)“will give satisfactory results”,参考译文将名词“results”置于前面,形容词置于后面,译为“效果较佳”,更符合汉语表达习惯。

[30] (1)原译“为测试本仪器”表述不完整;参考译文增译“的效果”,更清晰易懂。(2)原译将“preferably with a load”这一补充说明的信息(插入语)置于主要信息后,用逗号连

接,结构比较松散;参考译文将其置于括号内,信息主次更分明。(3)"并打开"这一表述欠妥。"并"不能体现操作步骤的先后顺序,且"打开"可能指"打开电压表"也可能指"打开探测器",容易使非专业人士误解。因此,参考译文将"and"译为"然后",并在"打开"后增译了"探测器(开关)"。

[31] 此处"back"是副词,但在此语境中跟上一个操作步骤之间有明显的动作过渡。因此汉译时将词性转换为动词。"well"是程度副词,表示"足够","stand back"表示"to move away from something, or to stand at a distance from something",可译为"退远站好"。"and"可省译,用语更简洁,结构更统一,更符合汉语表达操作流程的语言特点。

[32] (1)上文为"保持不动",下文立即出现"电压表应……",逻辑有些跳跃。因此,参考译文根据语篇衔接需要,增译了"测试过程中"。(2)"The meter"指的是上面提到的"voltmeter",汉译成"电压表"更清楚明了。(3)"indicate"后面名词较长时,汉译通常需要将名词的词性转换为动词或增译动词,因此参考译文增译了"有"字。"supply rail"可能有"电源轨""电源导轨""供电轨"等多种译法,这时将以上几种可能的译法分别与"电压表"一起检索,出现的结果基本都是"导轨"。

[33] "如果是这样"为增译内容,以保证上下文逻辑严密。前面是测试环节,电压表也可能是坏的,那电压就不能在一分钟后衰退为"0",因此"准备就绪"是有前提条件的。"decay"似乎可以译为"下降""减少""衰退""衰减"等,且网上均有此类用法。此时可以在搜索引擎检索"decay"电压,可发现"衰减"与其为专业搭配。

练习题

一、英译汉

Most engines are now equipped with an electronic ignition system (EIS), which is also called solid-state ignition (SSI). Instead of contact points, transistors and other semiconductor devices act as an electronic switch to turn the primary current on and off. With no contact points to wear, and with no need of adjustment or replacement, electronic ignition systems require less service. In addition, many electronic ignition systems produce higher secondary voltages than contact-point systems. This is important because higher secondary voltages permit engines to run on leaner air-fuel mixtures. Electronic ignition systems capable of producing higher secondary voltages are often called high-energy ignition (HEI) systems.

Higher secondary voltages require the use of spark-plug cables that have heavier insulation than cables used in contact-point systems. As a result, cables used in HEI systems are larger in diameter than cables used in contact-point systems. The solution is made of silicone compounds. They provide better insulating properties to contain higher voltages.

Some cars have an electronic distributor in which a magnetic-pickup coil is replaced by a sensor that uses the Hall effect. When a thin slice of semiconductor material carrying an electric current is crossed at right angles by a magnetic field, a voltage appears at the edges of the conductor. This voltage is called the Hall voltage.

The Hall voltage is proportional to both the current flowing through the conductor and the strength of the magnetic field. The voltage is unaffected by the speed at which the magnetic field cuts across the conductor.

The Hall-effect sensor is attached to the distributor. A small permanent magnet is also mounted on the distributor facing the sensor, with a small air gap between the sensor and the magnet. A distributor rotor is attached to the distributor shaft. The rotor has the same number of shutters (also called tabs or vanes) as the engine has cylinders. The gaps, or spaces between the shutters are called windows. As the shaft turns, the shutters and windows rotate through the air gap between the magnet and the sensor.

When no shutter (window) is in the air gap, the sensor senses the presence of a small magnetic field from the permanent magnet. As long as this magnetic field acts on the sensor, it sends a low voltage signal to the ECU. The voltage signal causes the ECU to keep the primary circuit open. No current flows through the ignition-coil primary winding. However, the instant the shutters enter the air gap, the effect of the magnetic field on the sensor is cut off. As a result, the Hall voltage drops to zero.

With no voltage signal from the sensor, the ECU closes the primary circuit. No current flows through the ignition-coil primary winding. This builds up a strong magnetic field around the coil. A current continues to flow through the primary winding as long as the shutters are between the magnet and the sensor. The instant the shutters move out of the air gap, the primary circuit is opened. This causes the magnetic field around the ignition coil to collapse. The resulting high-voltage surge is carried through the secondary circuit to the spark plug.

Dwell, or how long a current flows through the primary circuit, is determined by the width of each shutter. As the shutters enter the air gap, a primary current is turned on. When the shutters leave the air gap, a primary current is turned off.

The Hall-effect sensor eliminates the need for a centrifugal-advance mechanism in the distributor. The length and frequency of the voltage signals can be used by the ECU to provide a spark advance related to the engine speed.

二、汉译英

变速器设计者一直受到激励为机动车开发一个两踏板的控制系统,即加速踏板和制动踏板。这样的想法在现代的自动变速中已经有所体现。自动变速器简化了车辆的驾驶,而且对多数使用者而言,车辆运行得更平稳了。

目前,公路上行驶的汽车使用多种不同类型的变速器,有手动变速器、半自动变速器和自动变速器(包括无级变速器)。由于传统变速器难以实现无声平缓地、不间断动力地换挡,所以自动变速器通常采用啮合齿轮装置实现换挡,此装置中可通过多盘离合器和带式制动器选择不同速比,多盘离合器和带式制动器能以接合或断开传动系不同的组件来实现所需的各种速度。人们从汽车发明伊始就想使用自动或半自动变速系统来减轻驾驶疲劳。早期的机动车安装着笨重的变速箱,且换挡时易产生冲击。随后,许多汽车设计者开始研究一种更令人满意的变速方式。

自动变速器常称作全自动变速器,因此美国汽车工程师协会将其定义为“无人工辅助而自动有效地改变传动比的变速器”。自动变速器随负载改变而自动换挡,即在换挡时,动力持续输送到驱动轮。与半自动变速器不同的是,自动变速器换挡完全不用驾驶员操作,但它仍然允许驾驶员通过额外操作达到期许值。与手动变速器和半自动变速器相比,自动变速器自身的动力传输效率略低。尽管如此,在许多应用场合,可以用换挡系统,使发动机的运行维持在最经济的范围,从而弥补上述不足。

自动变速器的首度使用可追溯到20世纪早期的美国,时至今日,自动变速器的使用范围仍然很广,以致几乎没有多少美国人知道怎样手动换挡。自动变速器使驾驶轻松随意,尤其在车辆频繁启停的建筑密集区和城镇。挂上挡后,需要做的仅仅是踩油门踏板便走或踩制动踏板便停。

对载荷敏感的自动变速器所实现的功能就是接合传动系和选择速比,不需要驾驶员额外干预。每种自动变速器都包括两部分,前部包括替代了驾驶员操纵离合器的液力耦合器或变矩器,后部包括替代了标准的手动换挡操纵功能的阀系和液压控制单元。

自动变速器依据发动机的需求、不同的输入参数(发动机真空度、行驶速度、节气门位置等)选择挡位,保持最佳的动力。平时由离合器和手动变速器所完成的操作,可借助液力耦合器自动完成,液力耦合器允许在发动机和变速器之间小范围且可控地滑动。根据驾驶员的需要,微型液压阀(位于油门踏板处)控制不同速比,或者根据发动机条件和行驶速度进入设定的模式。

第十四章
交通运输工程类

【汉语原文】

不停车电子收费系统[1](节选)

不停车电子收费系统采用多种技术手段[2],使人工收费过程变成了自动收费过程[3]。如此一来,驾驶员不必在收费岗亭处停车用现金缴费[4]。有了不停车电子收费系统,连实体收费广场都已经不再需要了[5]。不停车电子收费系统设备安装在龙门架上[6],对车道内不同行驶速度的车辆进行收费[7]。

为使不停车电子收费系统能够高效、可靠地完成收费过程[8],通行能力达到最强,用户满意度达到最高,不停车电子收费系统就必须由三个主要部分组成[9],即车辆自动识别系统、车辆自动分类系统和逃费抓拍系统[10]。

车辆自动识别系统使用装备在车上的射频装置向收费装置传送识别信息,如 ID 号码、车型、车主等[11],以辨别车辆是否可以通过不停车收费车道[12]。车辆自动分类系统利用装在车道内和车道周围的各种传感器装置来测定车辆的类型[13],以便实现按照车型收费的正确性[14]。逃费抓拍系统用来抓拍不停车电子收费系统车道上未配备有效标识卡的汽车牌照[15],用于确定逃费车主并通知其应交费用或处罚办法[16]。所有这些系统都与我们熟知的"车道控制器"的装置相联结[17]。

车道控制器是一台计算机装置,接收车辆自动识别系统、车辆自动分类系统和逃费抓拍系统传来的信息[18]。通常每条车道仅安装一个车道控制器,与其他车道的设备配合使用,共同完成收费业务[19]。车道控制器也是维护有效标识卡清单的装置,用于确定由车辆自动识别系统提供的信息的有效性[20]。

除了这些车道内的设备外,每个收费广场通常都设有一台主计算机[21],收集车道控制器传送过来的业务信息,并且与采集各收费广场数据的收费管理中心主机交换信息,同时将数据存储到管理中心主机中[22]。收费广场主机的另外一个作用是向每个车道控制器传送有效标识卡清单,以确定车辆自动识别系统识别的有效性[23]。

最后,通常要设一处客服中心,为需要使用不停车电子收费系统的客户登记[24],管理客户交费账户,向客户发送标识卡,处理违章车辆抓拍图像,处理客户提出的问题等[25]。客户服务中心从安装在车道内的设备得到收费业务信息,并将信息计入相应客户的账户上[26]。

客户服务中心也向收费站的主计算机发送有效标识卡名单,然后再由主计算机将名单传给车道控制器,供车辆自动识别系统确定通过车辆标识卡的有效性[27]。

集成所有这些技术,是一项十分艰巨的任务[28]。多数部门都会委托供应商来将这些技术集成到现有收费系统中,或者开发一整套包括不停车收费在内的新收费系统[29]。

有了不停车电子收费系统,客户不再需要停车、开车窗、掏零钱,这样可以提高收费系统操作者的服务质量和客户的满意程度[30]。同时,客户如果选用不停车电子收费系统,还可以使用现金、支票甚至信用卡为缴费账户交费[31]。使用信用卡交费的客户常有这样的选择:交费账户余额低于预先规定的水平时,还可从信用卡上自动转账,因此免去了顾客为收费账户不断补充资金的烦琐事情[32]。此外,客户可以得到每月的详细交费清单,没有必要再去要收据[33]。

【文本分析】

本翻译文本节选自《不停车电子收费系统》,属纯信息类文本。通常该类文本的最大翻译单位为句子,最小翻译单位为搭配,翻译时应主要采用直译方法。但是原文本语言表述不够简洁且“学术腔”痕迹较为明显,个别地方还存在逻辑问题,因此应当以两句左右为最大翻译单位灵活处理。

原文主要用于科普,英译时应考虑目标受众的英语阅读水平,尽可能简洁明了,符合现代英语写作风格,立场客观、中立,但不应过于庄重。原文中有大量重复名词表达及不言而喻的信息(主要为名词),英译时应采取各种手段避免重复。个别地方涉及技术背景知识,译者应多方查阅、咨询,确保译文准确无误。

【原译】

Non-stop Electronic Charging System[1] (excerpt)

Non-stop electronic charging system adopts a variety of technical means[2] to make the manual toll collection process become an automated collection process[3]. In this way, the driver does not have to stop at the toll booth to pay the fee in cash[4]. With non-stop electronic charging system, even physical toll squares are no longer needed[5]. Non-stop charging system is installed on the air grid[6] to charge for vehicles with different speeds on the road[7].

To ensure the high efficiency and reliability of the electronic toll collection implementation process[8] and for the passage capacity to be the strongest and for the user satisfaction to be the highest, the non-stop charging system must be composed of three main components[9], namely the automatic vehicle recognition system, the automatic vehicle classification system and the toll evasion capture system[10].

The automatic vehicle recognition system uses an on-board radio frequency device to transmit identified information, such as ID number, vehicle type and owner to the toll collection device[11] to determine whether a vehicle can drive through the no-stop toll lane[12]. Automated vehicle classification systems use sensors installed in and around lanes to determine the type of vehicle[13] in order to achieve the correctness of toll by the type of vehicles[14]. The toll evasion capture system is used to capture the license plate image of the car that uses the toll lane but is not equipped with a valid identification card[15], so as to determine the car owner who avoids the toll and inform him of the fee or punishment method[16]. All of these systems are connected to devices commonly known as "lane controllers"[17].

The lane controller is a computer which receives information sent from the automatic vehicle identification system, the automatic vehicle classification system and the toll evasion capture system[18]. Usually, only one lane controller is installed on each and every lane. The controller works with devices of other lanes and complete toll collection services[19]. The lane controller is also a device that protects a list of valid identification cards used to determine the validity of the information provided by the automatic vehicle identification system[20].

In addition to these on-road devices, each toll square usually has a main computer[21] that collects the business information transmitted from the lane controller and exchanges the information with the toll management center mainframe that collects data from the toll plazas and stores the data in that main computer[22]. Another function of the toll plaza host is to transmit a list of valid identification cards to each lane controller to determine the validity of the automatic vehicle identification system[23].

Finally, a customer service center is usually set up to register customers who need to use the no-parking toll system[24], manage their toll accounts, send identification cards to customers, process images of illegal vehicles, deal with customer questions, etc[25]. The customer service center receives toll service information from the equipment installed in the lane and credits the information to the corresponding customer's account[26]. The customer service center also sends a list of valid identification cards to the main computer at the toll station, which then passes the list to the lane controller for the automatic Vehicle identification system to determine the validity of the passing vehicle identification card[27].

It is a daunting task to integrate all these technologies[28]. Most departments will entrust suppliers to integrate these technologies into existing toll systems or develop a whole new toll system that includes non-parking fees[29].

With a non-stop electronic charging system, customers no longer have to stop, open windows and pay for change, thus improving the quality of service for the toll system operator and customer satisfaction[30]. At the same time, customers can also use cash, checks or credit cards as payment accounts if they choose the non-stop electronic charging system[31]. Customers who pay by credit card often have the option of having their credit card automatically paid when the balance in their payment account is low, thus eliminating the need for customers to worry about paying the toll funds[32]. In addition, customers are provided with a detailed list of monthly payments and there is no need to ask for receipts[33].

【参考译文】

Electronic Toll Collection[1] (excerpt)

Electronic Toll Collection (ETC) is the use of various technologies[2] to automate the manual toll collection process in such a way[3] that customers do not have to stop and pay cash at a toll booth[4]. With ETC, an actual toll plaza is not even a requirement[5]. Mounted on overhead gantries[6], ETC allows vehicles to be charged while they proceed at highway speeds[7].

For an ETC implementation to be effective and reliable[8] and achieve maximum throughput and customer acceptance, three major systems are required[9]: Automatic Vehicle Identification (AVI), Automatic Vehicle Classification (AVC), and Video Enforcement Systems (VESs)[10].

AVI transmits, via a radio Frequency (RF) device installed in the vehicle, such identifying information as the ID number, vehicle type and owner to the toll equipment[11], to determine whether the vehicle can pass through the ETC lane[12]. AVC, with various sensors in and around the lane, determines the type of vehicles[13] so that the proper toll can be charged[14]. VESs capture images of the license plates of vehicles that use ETC without a valid tag[15] so that the owners can be identified and notified of a due toll or a penalty[16]. All these systems are connected by what is commonly known as a lane controller[17].

The lane controller is a computer device that receives inputs from the AVI, AVC, and VESs equipment[18]. Usually, there is, on each lane, one such controller that complete toll transactions in coordination with other lane equipment[19]. The lane controller is also the device that maintains a list of valid tags which it uses to validate the information provided by AVI[20].

In addition to all this in-lane equipment, each toll plaza usually has a host computer[21] that collects the transaction information from the lane controllers and communicates with a central host computer that collects and stores data from all plazas[22]. The plaza host computer is also used to transmit to each lane controller the list of valid tags used for AVI validation[23].

Finally, there is usually a Customer Service Center that enrolls customers[24], manages their toll account, issues tags, processes violation images, and handles customers'inquires[25]. This Center receives, from the in-lane equipment, toll transaction information and posts it to the appropriate customer account[26]. It also transmits to the plaza host computers the list of valid tags, which will, in turn, be sent to the lane controllers for use in AVI validation[27].

Putting all this technology together can be a daunting task[28]. Most agencies therefore hire a vendor, either to integrate all this technology into their existing toll environment or to develop a whole new toll system that includes ETC[29].

ETC allows the toll facility operator to improve customer service and satisfaction by removing the need for customers to stop, roll down their window, or fumble for change[30]. Customers have the flexibility of topping up their ETC toll account with cash, check, or even credit cards[31]. Credit card users may often opt to have their credit card account automatically charged in case of insufficient balance in their toll account, thereby eliminating their concern over funds for toll payment[32]. In addition, customers can receive monthly statements and will not have to ask for receipts[33].

【译文详解】

[1] 原译为字对字翻译。不停车电子收费系统(简称“电子收费”)最早在挪威推出,后兴起于美国,汉语名实则由英语翻译而来。

[2] (1)专有名词第一次出现时,应使用全称;如果下文多次使用,可在第一次使用时加括号备注简称。(2)“采用”一词虽可译为“adopts”,但“ETC”作为动作的施事主语,无法和“adopt”连用;参考译文做了词性转换,让名词主语和名词“use”之间的逻辑更为清楚。(3)“Technology/technological”主要指“modern scientific technology”,而“technique/technical”多指“useful or mechanic arts”,用于社会、学术、商务等层面。从这个角度来讲,此处宜用“technology”。此外,“technologies”一词用复数,就是表示具体的技术手段,因此参考译文省译了“手段”一词。

[3] 原译“toll collection process become an automated collection process”中出现了两次“process”和略显冗长的“make...become automated”表述,中式英语痕迹比较明显。参考译文中省译了一个“process”,用“automate”一个词替代原译中的“make ... become automated”,更为简洁明了。

[4] (1)原文第一、二句中的“自动化”和“不必停车……”存在逻辑关系,因此参考译文将两句合译为一句。(2)“用现金交费”就是“交现金”,因此参考译文省略了“pay the fee in cash”中的“the fee”,直接用更为简洁的“pay cash”。

[5] (1)“有”为汉语中的静态动词,英译时一般将这类动词转换为介词、名词、非谓语等,此处译为“with”最为直截了当。(2)原译中使用的“toll square”在BNC语料库中的使用频率为0,“收费广场”可在“术语在线”中查到相关表达“toll plaza”。(3)“physical”和“actual”均可与“toll plaza”搭配,表示“实体”。

[6] (1)“安装”和“收费”不是并列关系,可以按原译的理解方式处理为“is installed on the air grid to charge for vehicles”这种不定式结构,也可以用参考译文中的非谓语结构。参

考译文将"Mounted"提前,起到了强调作用。(2)"mount"和"install"的区别为"Mount is to physically attach an object to a wall or something else. Install means assemble something, to fix on to something and also to set up an application or software on to an electronic device"。换言之,"mount"多指将一个物体放置到另一物体上,而"install"涉及将零散物件组装成整体这一过程。因此此处的"安装"应译为"mount"。(3)在"术语在线"网站中查到"龙门架"又称"吊网门形架",对应英文表述为"gantry",是水产领域的专业词语。但经网页查询,交通领域中也有"龙门架"这一专有名词,用英语搜索引擎输入"ETC gantry"查看其图片,验证可以得知"gantry"为其正确译法。为使表意更加清晰,参考译文还增译了"overhead"一词,清楚描绘出了"龙门架"的状态。

[7] ETC费用的结算单位是银行,换言之,ETC本身并不"charge vehicles",而是银行通过ETC这一渠道去收费。基于这一事实,参考译文将原译的主动语态改为了被动语态。

[8] (1)原译将"高效、可靠地"这两个副词进行了词性转换,并将整句话译为"To ensure the high efficiency and reliability of the electronic toll collection implementation process"。但是,"并列名词 + of + 系列名词"这种结构可读性较差。原文简单来说就是"ETC收费过程高效、可靠",因此将副词转译为形容词,即"For an ETC implementation to be efficient and reliable",更为简洁易懂。(2)"过程"为可数名词,汉语量词用"次";而英语冠词"an implementation"就是指具体、可计算的"收费过程",因此参考译文省译了范畴词"process"。

[9] (1)汉语多为"Topic-Comment"(述评)结构,其显著特征是将形容词后置,原文"通行能力最强(大),满意度最高"便是如此。英语较短的定语通常置于名词前,因此初步译为"biggest throughput and maximum customer acceptance"。另外,该语境中的"biggest"和"maximum"同义。因此,参考译文将这部分处理为"achieve maximum throughput and customer acceptance",这比原译"for the passage capacity to be the strongest and for the user satisfaction to be the highest"更加简洁明了。(2)本句译文的前半部分已经出现了"ETC"这一信息,后半部分应避免重复,故参考译文省译了第二个"不停车电子收费系统"。

[10] (1)原文中提及的三个组成部分(车辆自动识别系统、车辆自动分类系统和逃费抓拍系统)均属于"systems",因此参考译文将"部分"处理为"systems",以避免术语中"system"一词的重复频率过高。(2)"recognition"和"identification"有所区别,"recognition"是指精确度较低的识别,"identification"指细微类别的识别。这两个词分别和"vehicle"一起用时,"vehicle recognition"仅表示能识别这个工具是车,而"vehicle identification"则表示能识别车的类别。(3)"逃费抓拍系统"这一说法是由英语翻译而来,英文原有的说法为"Video Enforcement Systems"。

[11] (1)原文有"使用"和"传送"两个动词,英译时应考虑将主要动词译为动词,次要动词

译为非谓语、介词、从句等结构。基于此,参考译文将"使用"的词性转换为介词,译为"via"。(2)"via a radio Frequency (RF) device installed in the vehicle"这部分可以放在句首、句中(类似插入语)或句末。具体位置要看信息的重要程度和语篇衔接,后者更重要。英文通常将最重要的信息放句末,次要信息放句首,其余信息放句中。因此,从信息的重要程度来看,"via"这部分放在句首更为合适。但是,考虑到上一段提到了"AVI""AVC""VESs"这三种系统,且本段末又再次提及,因此从语篇结构考虑,这段话的主语宜分别用"AVI""AVC""VESs"开头。(3)原译将"识别信息"译为"identified information",意思为"识别过的信息",但英语中这种用法很少;参考译文中的"identifying information"表示"用于识别某内容的信息",更符合原文的语境。(4)"on-board"仅表示"在车上"(有可能是随便放在车上任意位置的),未能体现"装备"的意思。从技术层面看,参考译文中的"installed in the vehicle"更为准确。(5)"such identifying information as..."和"identifying information, such as..."同义。前者少一个逗号,整体句式更加紧凑。

[12] "pass through"和"drive through"同义。虽然"vehicle drive through"的使用频率较高,但严格意义上,目前多数车辆尚不能自行"drive",因此参考译文用"pass through"更加严谨。

[13] 读者可以推知路边的 ETC 相关设备一般是安装好的,因此参考译文省译了"安装"一词。

[14] (1)中文的重复表达频率远高于英文,有时为了结构平衡、韵律节奏,中文倾向使用重复表达。原文"来测定车辆的类型,以便按照车型实现……"中连续出现两次"车型",中文可以接受,但英译时应省译后一个不言而喻的"车型"。(2)原文中"实现正确性"这一表述较为拗口,译为"achieve the correctness of"也比较刻意。这句话可简单理解为"实现正确收费",因此参考译文将其处理为"so that proper toll can be charged"。

[15] (1)逃费抓拍系统本身就可以抓拍,因此参考译文省译了"is used to"。(2)"plate images"在该语境中肯定是车的牌照信息,因此参考译文省译了"plate images of cars/vehicles"中"of cars/vehicles"这一不言而喻的信息。(3)原译"uses the toll lane but is not equipped with a valid identification card"没有语法错误,但不够简洁;参考译文将"未备有"的词性转换为介词,译为"uses the toll without a valid tag",简洁明了。(4)"有效标识卡"更常见的说法是"有效标签",应当译为"tag",而不是"card"。

[16] (1)"identify"比"determine"更常用于确定人的身份。(2)逃费抓拍系统并不能"确定"或"通知"车主,而是由其他系统或工作人员通知,因此参考译文将其处理为被动语态,使其逻辑显化,也更符合实际操作流程。(3)"逃费车主"中的"逃费"在该语境中不言而喻,故省译;可数名词"a penalty"暗含了具体的方法,因此参考译文省译了"方法"一词。

[17] (1)参考译文中的"what"相当于原译中的"device that",但"what"更为简洁、地道。(2)根据下文可知,这三个系统连接的只有一个车道控制器,因此"lane controller"应该使用单数。

[18] (1)车道控制器由工业控制计算机、串口扩展卡、I/O 接口扩展卡、视频捕捉卡、外设接口控制盒、配线架、电源和设备机箱等组成,其核心是computer。原译省译了"设备",意思稍有偏差;参考译文保留了"device",更加准确。(2)"from"和"to"都是方向感比较明显的介词,汉译英时常用来替代"来""送"等动词,因此参考译文将"传来的信息"转译为"information from"。

[19] (1)本句原文逗号前后的内容存在明显逻辑关系,因此译为一句话更为恰当。(2)英语写作提倡定语从句尽量紧靠先行词。本句如果写为"there is one such controller on each lane that",则 that 从句直接修饰的是"lane"。因此,参考译文将 that 引导的从句置于谓语与宾语之间,作插入语;这句话也可以写为"Usually, on each lane there is only one such controller that..."。(3)原译将"配合使用"(works with devices of other lanes)和"共同完成业务"(complete toll collection services)处理为并列关系,使其逻辑模糊不清。参考译文厘清了两者之间的主次关系,将"配合使用"的词性转换为介词结构,突显"完成收费业务"这一动作。(4)原译用了"complete toll services",但这种搭配十分少见。另外"收费业务"译为"toll services"也不全面;"收费业务"不仅包括收费,还包括图像信息存档等,应译为"toll transaction"(a record of activity created by the toll collection system as a result of a vehicle traveling through a tolling point)。

[20] (1)原译将"维护……清单"译为"protect...list",容易让人产生误会。这里的"维护"应为"maintain"(keep something in existence at the same level, standard, etc.)。(2)维护这个清单的装置就是这个车道控制器,此处应该为特指,故将原译的"a"改为"the"。(3)原译"determine the validity of something"不够简洁,用"validate something"即可将意思表达清楚。

[21] 原译将"车道"译为"road",在该语境中不够具体;"车道"一般译为"lane"。"主计算机"应译为"host computer",意为"a computer or server connected to a network and providing facilities to other computers and their users"(https://www.collinsdictionary.com/dictionary/english/host-computer)。

[22] (1)参考译文省略了"transmitted",其原因请参阅本章[18](2)。(2)"主机"就是"主计算机"的简称,即"host computer";原译将"主机"误译为"framework"。原文有两个"收费管理中心主机"(host computer),英译时只需保留一个,然后用"collects"和"stores"两个动词并列加上"host computer"这一宾语,更加简洁明了。(3)"管理中心主机"译为"host computer in the toll management center"不够简洁,且这几个词后面的 that 从句在语法上修饰"management center",这就会导致信息错误。参考译文根据行

业习惯,将“管理中心点主计算机”译为“central host computer”,简洁且无歧义。

[23] 原译虽然省译了“识别的”,但表达仍然不够简洁。参考译文“transmit to each lane the list of valid tags used for”简洁地道。其中,“to each lane”的位置只能放在“transmit”后,类似于插入语类。如果写成“to transmit the list of valid tags to each lane controller used for”,其中的“used for”修饰的是“controller”,这就偏离了原文意义。

[24] (1)原译将“设”这一非重要动词译为“is set up to...”,体现的信息层次不够清晰;参考译文将“设”处理为 be 动词,将后面的“enroll”等动词突显出来。(2)该语境中,不停车收费系统客服中心肯定是为 ETC 客户提供服务的,因此参考译文省略了不言而喻的“who need to use ETC”。(3)“register”和“enroll”都有“登记”之义。但“register”只包含“signing up”这一步,“enroll”包含了“register”及其后的缴费环节。因此,参考译文选用了“enroll”一词。

[25] (1)原译将“发放”处理为“send”,信息错误。“send”一词还可能指“寄送”,而原文的重点是“发送”,即“issue”。(2)原译将“违章”译为“illegal”(违法),属于过度翻译。参考译文中的“violate”在该语境中指“违章”。另外,“violation images”暗含了“vehicles”之意,因此参考译文省译了“车辆”一词。(3)“deal with questions”搭配不当,常用搭配为“address/answer/respond to questions”。此外,“question”还有“质疑”之义,在该语境中可能引起不必要的误会,因此参考译文选用了更为客观的“inquiry”一词。

[26] (1)上一句的主语为“Customer Service Center”,参考译文中这句用“This Center”避免重复。(2)若将本句译为“This Center receives...information from...equipment and post it”,则“it”在语法上指代的是“equipment”。虽然专业人士清楚“it”实则指代“information”,但这并不是最清楚易懂的译法。

[27] (1)参考译文将“to the plaza host computer”与“the list of valid tags”调换位置,能避免错误修饰。(2)“传”是“传送”,译为“send”。(3)根据前一段最后一句话的表述,参考译文将“供车辆自动识别系统确定通过车辆标识卡的有效性”直接简化为“(for use) in AVI validation”,体现了翻译过程中的语篇衔接意识。

[28] 原译欠妥。其一,“集成所有这些技术”本身很短,没必要增加形式主语,使句子变长。其二,前几段一直在讲技术,这一段将“technology”后置对语篇衔接不利。参考译文则将“technology”这一旧信息提前,“daunting task”这一新信息置于句末,更符合认知习惯。

[29] (1)原译中“entrust”的搭配不当;“entrust”的搭配多为“entrust something/somebody <u>to somebody</u>”或“entrust B with A”。另外,原文提到供应商还可能来开发新系统,其中必定会涉及收费问题。因此,参考译文用了更符合该语境的“hire”一词。(2)目的状语中的两个宾语都比较长,一个宾语前有 V-ing 形式的形容词,另一个为宾语从句,因此

参考译文采用"either to do or to do"结构,使整个句子结构更加平衡、层次更加分明。

[30] (1)在不影响语流的情况下,英语更常将结果放句首,原因放句末,表结果的内容比较短时更是如此。原文中"提高服务质量"和"提高满意度"两个信息属于结果,且相对较短,因此参考译文中的"ETC allows...to improve customer service and satisfaction by removing the need..."比"ETC removes the need...to improve..."更符合英语行文习惯。(2)原译将"有了某某某,某人/事如何"这一句型译为"介词短语 + 主句"结构,不够简练,逻辑稍差。当该句型中的"某某某"跟后面的"某人/某事"存在明显因果关系时,可以省译介词,增译动词,显化其中的逻辑关系。因此参考译文将"With ETC, customers no longer need"改为"ETC allows..."。(3)车不能从"booth"里穿过去,故"收费站"不应译为"toll booth"。"开车窗"译为"roll down the window"(摇下车窗)更具体。"掏零钱"暗含拿出合适的现金金额很麻烦之意,用"fumble"更形象;原译中的"pay for change"为低级错误。(4)交费时只需要开驾驶员所在的那一扇车窗,因此"window"用单数。

[31] (1)原文存在逻辑问题:"还"跟上下文没有任何联系,应当删去。(2)"为缴费账户交费",这里的交费类似于充值(top up an account)。

[32] (1)原译"Customers who pay (top up) by credit card"不够简洁,参考译文根据语境将其简化为"credit card users"。(2)原译中"have credit cards paid"存在逻辑错误,这里指的是银行系统"charge"ETC 用户。(3)原译"when the balance in their payment account is low"部分不够简洁;参考译文用介词结构"in case of insufficient balance"更简洁且句子层次更清晰。(4)原译"the need for customers to worry about"比较冗余;参考译文用"their concern over"更简洁。(5)"customers can receive"比"customers are provided with"更简洁,可读性更强。

[33] (1)"每月的详细交费清单"中的"交费清单"根据英美国家惯例译为"statement"而非"list"。同时,百科网站中"monthly statement"的解释为"A monthly statement is a written record prepared by a financial institution, usually once a month, <u>listing all credit card transactions</u> for an account, including purchases, payments, fees and finance charges"。因此,该术语已经包含"所有交易明细"之义,因此参考译文省译了"详细"。(2)原文"没必要……"这部分的主语就是"客户",一个主语接并列谓语最为简洁;原译的"there is no need to do"为这句话增加了一个句型,可读性较差。

练习题

一、英译汉

At urban road junctions it is often not possible to provide full standards for large vehicles. Where their numbers are small and speeds are low, they can be allowed to dominate two or more traffic lanes at urban roundabouts and traffic signals. Traffic islands, kerb lines and street furniture must, of course, be placed outside the vehicle swept paths. Standard vehicle swept path plots are available for checking layouts but for more complex multiple turns the DoT computer program TRACK is recommended.

Apart from abnormal load vehicles the longest vehicle in everyday use in the UK is the 18m drawbar trailer combination. When designing junctions at confined sites, it is not enough to check that the layout is adequate for the largest vehicles. Swept path is dependent on a number of factors: rigid or articulated; wheelbase (or tractor wheelbase); front and rear overhang; width; and length of trailer.

In some instances an articulated vehicle has a smaller turning circle and smaller swept areas than a long rigid vehicle. Vehicles with long front or rear overhangs, such as the 12m European low floor bus can pose particular problems. If such a vehicle is stationary and its steering wheels are on full lock, the rear will tend to move outwards as it starts to move. If the vehicle is close to the kerb, the bodywork might strike a pedestrian.

The selection of junction type can be very simple and obvious in some cases. For example:

- two lightly trafficked residential roads—priority junction.
- the through carriageway of a motorway—grade separation.
- heavily trafficked urban crossroads with heavy pedestrian flows—traffic signals.
- suburban dual carriage ways with substantial heavy goods traffic—conventional roundabout.

However, there are very many cases where the solution is anything but obvious and the engineer should resist making a decision until all the evidence has been examined and analyzed.

Frequently, when an existing junction is to be upgraded to handle more or different types of vehicles, the existing form or control method must be considered. The form of the other junctions on the main route might determine the form of the new or improved junction; for example, it might be inappropriate to construct a mini-roundabout on a route that has a series of linked signaled junctions and pelican crossings. Similarly, a signaled junction on an otherwise free-flowing rural dual carriageway with generously designed priority junctions or roundabouts could prove to be hazardous.

Priority junctions are the simplest and most common of intersections and the control at priority junctions depends upon give-way road markings and post mounted signs.

Crossroad layouts should be avoided wherever possible because they concentrate a large number of vehicle movements and, therefore, conflicts within the junction. On unlit rural roads at night, drivers on the side roads can be confused into thinking that they are on the major route, particularly when there is an approaching vehicle on the far side of the crossroads.

二、汉译英

讨论视距要分四步:用于停车要求的距离(适用于所有的道路);超越被超越车辆所需要的距离(仅对两车道道路适用);在复杂地点进行决策所需要的距离;用于设计的测量距离的标准。

停车视距。视距是驾驶员可见的前方道路长度。道路上应有的最小视距应该足够长,这样如果车辆以等于或接近设计速度行驶,才能在撞到行驶路线上的障碍物之前停下来。虽然说视距越长对行车越有利,但沿道路每一点的视距至少应能使低于平均操作水平的驾驶员或平均性能的车辆在这个距离内停车。

停车视距是两个距离之和:从驾驶员发现有必要使其停车的障碍物到制动开始这期间所走的距离,加上从制动开始到停车所走过的距离。这两个距离分别叫作制动反应距离和制动距离。

设计视距。停车视距的长度通常能够满足具有一般操作能力和警觉意识的驾驶员在正常的情况下停车。但是,当驾驶员必须作出复杂或瞬间的判断,或难以感知某些信息或需要采取特殊操作时,上述停车距离往往是不够的。停车视距可能不会为驾驶员提供足够的能见距离,使驾驶员能够采取提前警报或执行必要的操作。显然,很多地点需要设计较长的视距,这样才是比较谨慎的做法。在这些情况下,采用决策视距能为驾驶员提供所需的较长的距离。

决策视距是指:驾驶员在可能引起视觉混乱的道路环境中发觉一个意外的、难以察觉的信息源或危险,识别危险或其潜在威胁,从而选择适当的速度和路线,并妥善、有效地完成安全操作所需要的距离。由于决策视距提供驾驶员避免失误的额外保障,它可以使得驾驶员在匀速或减速的情况下操控车辆,而非必须停车,所以它的取值比停车视距要大得多。

两车道的超车视距。大多数道路和很多街道都属于双向两车道道路,在这类道路上行驶的车辆会频繁超越前方速度较慢的车辆。通常,超车行为必须利用反向车道来完成的。如果想安全完成超车过程,驾驶员必须能够看见前方足够远的距离且确保道路畅通无阻,同时,还要保证超车行为是在遇到对向来车前对被超车辆的正常运行无任何干扰的情况下完成。当超车行为只完成一部分,这时该车与对向来车已经很近,驾驶员则有必要退回原车道。许多超车行为都是驾驶员在没有看清楚前方安全距离的情况下完成的,基于这种情况的设计并不能达到理想的安全系数。由于有许多较为谨慎小心的驾驶员不会在这种情况下超车,所以,以此为基础的设计会降低道路的使用率。

多车道视距。在每一行驶方向上具有两车道及以上的道路或街道上没有必要考虑超车视距。多车道道路的超车行为一般在每条单行道的控制线内完成。因此,跨越无分隔设施的四车道道路中线或者跨越四车道道路中央分隔带的超车行为属于鲁莽驾驶,应当予以禁止。

第十五章
土木工程类

【英语原文】

HARD ROCK[1]

Ultimate Bearing Capacity[2]

The above soil properties of the earth will be measured by the Contractor at the various locations in conformity with the standard method of testing[3] and the foundation design will be revised suiting the site conditions from such tests[4].

Properties of Concrete[5]

The cement concrete used for the foundations shall generally be of grade M-20[6] having 1:1.5:3 nominal volumetric mix ratio with 20mm coarse aggregate for chimney portion and 20mm/40mm aggregates for pyramid or slab portion[7]. All the properties of concrete regarding its strength under compression, tension, shear, punching and bending etc. as well as workmanship will conform to IS:456[8].

The parameters of concrete to be considered for design of foundations are given in TABLE 2-2[9].

Parameters of concrete for design of foundations Table 2-2

Parameters	Value
WEIGHT OF CONCRETE	…
TYPE OF CONCRETE	…
WEIGHT OF DRY REGION kN/m^3 (kg/m^3) [10]	…
WEIGHT IN PRESENCE OF SUB-SOIL WATER kN/m^3 (kg/m^3)	…
Plain Concrete	…
Reinforced Concrete[11]	…

The Quantity of minimum cement to be used per unit quantity of consumption for different mix (nominal mix) of concrete should be as follows[12]:

Quantity of Minimum cement to be used per Unit quantity of work (in kgs)(...) [13];

1:1.5:3 nominal mix concrete;

Random Rubble Masonry with 1∶6 cement mortar [14]。

In this regard utilisation record is to be maintained at site[15].

Alternatively, Ready Mix concrete from batching plant as per IS 4925 can also be used with no extra payment and without any recovery[16]. However, Cement content shall be as per IS 456. The ready mix concrete shall conform to IS:4926[17]. The selection and use of Materials for the ready mix concrete shall be in accordance with IS:456[18]. The concrete shall be of M20 grade design mix as per IS:456. The transport of concrete and transportation time shall be as per IS:4926[19].

Cement used shall be ordinary Portland Cement, unless mentioned otherwise, conforming to the latest Indian Standard Code IS:269 or IS:8112 or IS:12269[20].

Alternatively, other varieties of cement other than ordinary Portland Cement such as Portland Pozzolana Cement conforming to IS:1489 (latest edition) or Portland Slag Cement conforming to IS:455 (Latest edition) can also be used. The Contractor shall submit the manufacturer's certificate, for each consignment of cement procured, to the Employer[21]. However, Employer reserves the right to direct the Contractor to conduct tests for each batch/lot of cement used by the Contractor and Contractor will conduct those tests free of cost at the laboratory so directed by the Employer[22]. The Contractor shall also have no claim towards suspension of work due to time taken in conducting tests in the laboratory[23]. Changing of brand or type of cement within the same structure shall not be permitted without the prior approval of the Employer[24]. Sulphate Resistant Cement shall be used if sulphate content is more than the limits specified in IS:456, as per Geotechnical investigation report[25].

The curing time of cement will be decided at the time of execution of the work under the contract based on the certificate from a reputed laboratory which will be obtained and submitted by the Contractor[26].

Concrete aggregates shall conform to IS:383-1970[27].

The water used for mixing concrete shall be fresh, clean and free from oil, acids and alkalis, organic materials or other deleterious substances. Potable water is generally preferred[28].

Reinforcement shall conform to IS:432-1966 for M. S bars and hard drawn steel wires and to IS:1138-1966 and IS:1786-1966 for deformed and cold twisted bars respectively[29]. Contractor shall supply, fabricate and place reinforcement to shapes and dimensions as indicated or as required to carry out intent of drawings and specifications[30].

Measurement, Unit Rates and Payment for Foundation

The indicative shape of foundations is enclosed in this Specification[31]. The bidder is required to quote the unit rates for different foundation types for a particular tower in the relevant Price Schedule[32].

The Bidder has to provide in the Bid the guaranteed foundation quantities (i. e., Excavation volume, Concrete volumes and Weight of Reinforcements) and unit rates for excavation, concreting and reinforcement for each type of foundation (as classified in clause 2.2 of this section) for each type of tower[33]. Composite price quoted (as described in clause 3.4 of this section) in respective Schedule for each type of foundation must comply with unit rate quoted and guaranteed foundation quantities mentioned[34].

The concrete volume and dimensions of the foundation shall be determined from the drawing approved. Measurement of concrete volume shall be in cubic meters and shall be worked out to the second place of decimal[35].

The steel required for reinforcement of foundation shall be provided by the Contractor[36]. Measurement will be based on the calculated weights of actually used in tonnes corrected to third place of decimal, no allowance being made for wastage[37]. No payments will be made for wire required for binding the reinforcement, chairs, bolsters and spacers, as the cost of these is deemed to be included in the unit rate quoted for the item of reinforcement[38].

【文本分析】

原文本节选自某《土建施工方案细则》,工程技术专业知识含量高,术语规范,整体表述流畅。科技类文本多属于信息类文本(informative texts),翻译时应保持中立、客观,最大翻译单位为句子,最小翻译单位为搭配。基于以上分析,本文本的翻译重难点为长句的拆译和技术词语的处理,个别地方为了符合汉语搭配习惯需要适当增词或减词。另外,翻译时,建筑领域的度量单位应当尽量译出。

【原译】

坚硬的岩石[1]

最终承载力[2]

上述地基土的特性将由承包商将按照标准测试方法在不同地点加以测试[3],并根据这些测试结果修改地基设计,以适应现场条件[4]。

混凝土的性能[5]

用于地基的砂浆一般必须为M-20级[6],标称容积混合比为1:1.5:3,烟囱部分为20mm粗骨料,金字塔或楼板部分为20mm/40mm骨料[7]。混凝土在压缩、拉伸、剪切、冲压和弯曲等方面的强度以及工艺的所有特性都应符合IS:456的规定[8]。

表2-2中给出了设计地基时需要考虑的混凝土参数[9]。

基础设计中的混凝土参数 表2-2

参数	指标
混凝土重量	略
混凝土类型	略
干燥区域的表观密度 千牛/立方米(千克/立方米)	略
存在地下水时的重量 千牛/立方米(千克/立方米)[10]	略
素混凝土	略
钢筋混凝土[11]	略

不同混凝土配合料(标称配合比)每单位消耗量使用的最小水泥用量应如下[12]:

每单位工作量使用的最小水泥用量(kg)(略)[13];

1:1.5:3 的标称配合比混凝土;

1:6 水泥砂浆乱石墙[14]。

在这方面,应在现场保存使用记录[15]。

或者,根据国际标准 IS 4925,混凝土搅拌站产出的商品混凝土也可不另付款使用、零回收[16]。但水泥含量应遵循国际标准 IS 456,商品混凝土须符合国际标准 IS:4926[17]。商品混凝土的材料的选择以及使用必须遵循国际标准 IS:456[18]。根据国际标准 IS:4926,混凝土的设计配合比应当为 M20。混凝土的运输及运输时间必须符合 IS:4926 的规定[19]。

除另有说明外,使用的水泥应为普通波特兰水泥(硅酸盐水泥),以符合最新《国标准代码》IS:269 或 IS:8112 或 IS:12269 的规定[20]。

或者,也可以使用除普通硅酸盐水泥以外的其他水泥品种,如:火山灰硅酸盐水泥[据 IS:1489(最新版)]、矿渣硅酸盐水泥[据 IS:455(最新版)]。承包商必须向业主提交每批采购水泥的制造商证书[21]。然而,业主对指示承包商对承包商使用的每批/大量水泥进行试验保留权利,承包商将在业主指示的实验室免费进行这些试验[22]。承包商也不得因在实验室进行试验所花费的时间而暂停工作[23]。未经业主事先批准,不得在同一结构内改变水泥的品牌或类型[24]。根据岩土工程勘察报告,如果硫酸盐含量超过 IS:456 中规定的限值,则应使用抗硫酸盐水泥[25]。

水泥的固化时间将在合同规定的工程实施时根据承包商获得并提交的知名实验室的证书来决定[26]。

混凝土骨料应符合 IS:383-1970 标准[27]。

用于搅拌混凝土的水应是新鲜的、干净的、不含油、酸和碱、有机物或其他有害物质。一般来说,饮用水是首选[28]。

钢筋应符合 IS:432-1966 的 M. S 棒和硬拉钢丝,以及 IS:1138-1966 和 IS:1786-1966 的变形和冷扭曲棒[29]。承包商须提供、制作和放置钢筋,其形状和尺寸如图所示,以执行图纸及说明书的意图[30]。

地基的测量、单位价格和付款

地基的指示性形状附在本规范中[31]。投标人需要在相关的价格表中对特定塔楼的不同基础类型的单价进行报价[32]。

投标人必须在投标书中提供每种类型的塔的保证地基数量(即挖掘量、混凝土量和钢筋重量)以及挖掘、混凝土和钢筋的单价(据本节第2.2条规定)[33]。每种类型的地基在各自的附表中所报的综合价格(据本节3.4条所述)必须符合所报的单价和所述的保证地基数量[34]。

基础的混凝土体积和尺寸应根据批准的图纸确定。混凝土体积的计量单位必须为"立方米",并计算至小数点后第二位[35]。

地基加固所用钢筋须由承包商提供[36]。测量按照实际使用吨数(修改到小数点后3位)的计算重量进行,没有浪费补贴[37]。无须为固定钢筋、椅子、垫子和垫片所需的电线支付费用,因为这些费用被视为包括在钢筋项目报价中[38]。

【参考译文】

硬　　岩[1]

地基土极限承载力[2]

承包商将在不同地点依照标准测量方法测试上述地基土特性[3]。根据测试结果,由专人修改基础设计图,以使基础适应现场地质条件[4]。

混凝土性能[5]

基础所用水泥混凝土一般应当为M20级,标称体积下的配合比为1∶1.5∶3[6]。立柱部分为20毫米粒径的粗骨料,锥形部分或平板部分的骨料粒径为20/40毫米[7]。混凝土的抗压强度、受拉强度、抗剪强度、冲切强度、抗弯强度等所有特性以及工艺都要符合印度标准(印标)IS 456[8]。

设计基础所需考虑的混凝土参数见表2-2[9]。

基础设计中的混凝土参数　　表2-2

参数	指标
混凝土重量	略
混凝土类型	略
干燥区域的表观密度　千牛/立方米(千克/立方米)	略
存在地下水时的重量　千牛/立方米(千克/立方米)[10]	略
素混凝土	略
钢筋混凝土[11]	略

不同混凝土混合料(标称配合比)中,每单位耗量的最小水泥用量应如下[12]:

每单位工程量的最小水泥用量(以千克计)(略)[13];

混凝土标称配合比为1:1.5:3;

碎石砌体水泥砂浆配合比为1:6[14]。

以上用量,应当当场记录[15]。

也可以使用符合印标IS 4925规定的配料厂生产的商品混凝土,不需要额外付费;没用完的,不予回收[16]。但商品混凝土的水泥用量应当符合印标IS 456,商品混凝土应当符合印标IS 4926[17]。选择、使用商品混凝土材料时,应当遵循印标IS 456[18]。依据IS:4926的规定,混凝土的设计配合比应当为M20。混凝土运输方式及运输时间应当符合印标IS:4926的规定[19]。

所用水泥应当为符合最新印标规范IS 269、IS 8112或IS 12269的普通波特兰水泥(普通硅酸盐水泥),另有规定的除外[20]。

此外,也可以使用除普通硅酸盐水泥以外的其他水泥品种,如符合印标IS 1489(最新版)的波特兰硅酸盐水泥或符合印标IS 455(最新版)的矿渣硅酸盐水泥。对每批已购水泥,承包商均应当向业主提交厂商证明文件[21]。但雇主有权指示承包商检测其使用的每批水泥[22],承包商按雇主的指示,到实验室免费参加检验,但不得因实验室检验耗时而中止工作[23]。未经雇主提前批准,不得在同一结构中更换水泥品牌或种类[24]。根据岩土工程勘察报告,工程岩土硫酸含量超过IS 456规定限值的,应当使用抗硫酸盐水泥[25]。

水泥养护期。承包商应先从知名实验室获得并提交水泥检验证明,根据证明在合同约定的动工时间决定水泥养护期[26]。

混凝土骨料应当符合印标IS 383 (1970)[27]。

用于搅拌混凝土的水应当是干净的淡水,不含油、酸、碱、有机材料等有害物质。一般首选饮用水[28]。

低碳钢筋和硬拉钢丝应当符合印标IS:432-1966,变形钢筋及冷轧扭钢筋应当符合IS:1138-1966及IS:1786-1966两个印标[29]。承包商应当按照所需形状和尺寸,提供、制作、安装钢筋,以满足施工图纸和施工细则的要求[30]。

地基基础的测量、单位价格及付款

所需地基基础的形状附在本《细则》中[31]。投标人需要根据相关《投标报价表》填报具体大楼不同基础类型的单价[32]。

投标人必须在投标中写明各类大楼各类基础(按本节第二条第二款分类)的保障工程量(即:挖方量、混凝土体积、钢筋重量)及开挖单价、浇筑混凝土单价和钢筋单价[33]。各类基础各附表中的合价(如本节第三条第四款所述)必须与先前提到的单价和基础保障工程量的乘积一致[34]。

混凝土体积和基础尺寸应当根据获批的图纸而定。混凝土体积应当用“立方米”计量,计算至小数点后第二位[35]。

基础加固所需的钢材应当由承包商提供[36]。按实际使用的计算重量(以吨为单位)计量,保留到小数点后第三位,不考虑浪费[37]。钢筋绑扎丝、脚手板绑扎丝、承枕绑扎丝、垫块绑扎丝视为包含在钢筋项目的报价单价中,不再付款[38]。

【译文详解】

[1] 原译“坚硬的岩石”听起来不够专业,业内多用“硬岩”。

[2] 原译为“最终承载力”,让人不禁质问“土壤还有最初承载力”吗？根据词典定义,“ultimate”有“most extreme”之意,因此译为“极限承载力”更为恰当。另外,根据上下文推断,此处的土壤为“地基土”,增译出来后意思更清楚明了。

[3] 汉语多重状语的一般顺序是目的、时间、地点、条件、比况、方式、伴随、频率、指涉(对象)、程度。此处的“依照标准测量方法”是方式,应该放在“不同地点”之后。节选部分前面还有一部分内容是“soil properties”(岩土特性),而原文将“soil properties”放在句首,使用被动语态,是为了将旧信息置于句首,方便衔接。汉语是单音节词,不像英语那样占空间,将“上述地基土特性”置于句末并不影响阅读。

[4] (1)“and”一词可以省略,汉译时另起一句话。(2)“foundation”一般译为“基础”,基础是房屋结构的地下部分;地基是承载所有重力的土或岩石,不是房屋建筑的一部分。“foundation design”在房屋设计中指的基础设计,且修改一般搭配“图”或“图纸”,故译为“基础设计图”。(3)“现场条件”比较笼统,因此根据实际情况增译“地质”一词。设计(图)本身并不能适应地质条件,而是设计的建筑基础适应条件,故增译“基础”一词。(4)“will be revised”的原译给读者的感觉是承包商去修改图纸,这并不符合实际分工情况,因此参考译文中增译模糊指代的“专人”。

[5] 作小标题时,省略“的”字,更为简洁。

[6] (1)“cement concrete”直译为“水泥混凝土”,而原译为“水泥砂浆”。“水泥混凝土”是水泥、砂、石等用水混合结成整体的工程复合材料的统称。“水泥”“混凝土”“砂浆”三者有所区别。“砂浆”中不含“石子”。而通过 Quora 搜索得知,“cement concrete”中含有“coarse aggregate”(骨料)。综上,“cement concrete”仍应译为“水泥混凝土”。(2)“shall”在法律英语中十分常见,本文兼具法律英语特点,法律界一般将其译作“应当”,而非“必须”。(3)“mix ratio”原译为“混合比”,网络释义为“混合比例”“混合比”“配合比”“配比”,可将这几个词置于“混凝土”之后,依次在语料库中搜索,很容易确定“配合比”更为合适。(4)“nominal volume”一般译为“公称容积”,但“容积”一般用于能装东西的物体,此处应为“体积”。

[7] “coarse aggregate”可能译为“粗集料”“粗骨料”等。通过百科知识网站查询可知,骨料是指支持混凝土的粗集料,又名粗骨料,集料一般来说分为 0 ~ 4.75mm 的细集料与大于这个范围的粗集料(骨料),而细集料的作用并不是支撑。原译“20mm 粗骨料”不够明确。查询得知,20mm 为粒径,故参考译文中增译了“粒径”一词。原译将“chimney”

译为“烟囱”,“pyramid”译为“金字塔”,“slab”译为“楼板”,明显有误。如果是整栋楼,出现这些词则完全可以理解,但是地基基础是埋在下地的,没有金字塔、楼板等部分。检索“chimney”(地基基础)可找到“立柱”一词,在“知识贝壳”中检索得知“slab”为“平板”,“pyramid”为“锥形(部分)”。

[8] (1)英语忌讳重复,因此用一个“strength”统领后面所有的特性;中文对重复词语的包容性较强,汉译时应将各种强度的专业名词悉数写出。(2)“workmanship”与“properties”为并列成分。“as well as”的用法与“and”有所差异,详情请参阅 https://www.differencebetween.com/difference-between-and-and-as-well-as/。(3)原译增加了“的规定”几个字,使其更符合汉语搭配习惯。参考译文增译了“印度标准(文件号)”,也是出于对词语搭配的考虑。经查源文件,本翻译练习里提到的“IS”文件均指印度标准而非国际标准。某一标准首次出现时,应当使用全称,下文可以简称为“印标”。

[9] 原译采用了逆序法,将“表2-2”置于句首。单独翻译这句时,这种译法并无不妥之处,但考虑到上下文衔接问题,将“基础”“混凝土”等前文已提及的信息放在句首更为恰当。

[10] 通过术语搜索得知“kN/m^3”这一相对生僻概念过去译为“容重”,现行技术标准中多统一译作“表观密度”。同时,还能查到“容重”的另一个单位是 kg/m^3,这也再次证明两个单位指向同一个概念。

[11] “concrete”译为“混凝土”,但是要寻找“plain concrete”对应的汉语术语可以在搜索引擎中输入“plain 混凝土”,很容易查询到参考译文中的“素混凝土”。以同样的方法,输入“reinforced 混凝土”,查询到“钢筋混凝土”。

[12] “mix (nominal mix) of concrete”的行业惯用译法为“混合料”,而不是“配合料”。混凝土需要按一定比例配制,因此该过程称作“配合”,而不是“混合”。配比完成后,需要将各种材料混合,这个过程一般需要搅拌,搅拌后的成品叫“混合料”。

[13] “work”原译为“工作量”,为符合建筑行业语言表述,将其改译为“工程量”更为恰当。

[14] (1)上文提到的主题是“quantity”(用量),那紧接着的下文重点应突出“量”;中文突出数量时,数字通常置于句末,如“发表论文30余篇”比“发表30余篇论文”更常见、更符合汉语表达习惯。基于此,参考译文将具体比例后置,同时增加“为”字,使其符合汉语语法习惯。(2)“Random Rubble Masonry”原译“乱石墙”来自机器翻译,和原文学科背景缺少联系。通过检索土木工程类术语网站可以确认该术语应译作“碎石砌体”。

[15] 原译“现场保存使用记录”令人费解,在工程英语中“maintain records”多指“做记录”;“at site”译为“当场”更地道。“In this regard”的原译比较笼统,此处需要将信息具体化。

[16] (1)“as per”有“in accordance with”或“according to”的意思,原译就理所当然地将其译为“根据……(规定)”。通过查询 IS 4925 源文件可知,这个标准文件是对“ready mix batch plant”作了细致的规定,却并未提到“ready mix”本身。由此推断,这个标准文件对用不用“ready mix”无权限制。综上,“Ready Mix concrete from batching plant as per IS

4925”是指“符合 IS 4925 规定的配料厂”。(2)原译“搅拌站”多为机器翻译结果,“搅拌站”的英语是“mixing plant”;通过查询术语网站及百科网站,“配料厂”“配料机”“搅拌站”基本一致。在 Quora 里搜索“batching plant vs mixing plant”,发现“batching”更多是用来配料,“mixing”多是搅拌或者拌和。综上,译为“配料站”(2013 年公布的机械工程名词)更为稳妥。(3)“生产”为增译而来。(4)“零回收”不太容易理解,通过结合土木工程专业知识,需增译作“没用完的”。

[17] 原译“水泥含量”不够准确,改译为“水泥用量”。根据上下文增译“商品混凝土的”。

[18] 在科技汉语欧化的背景下,原译“商品混凝土的材料的选择以及使用”在句式上无明显问题,但是用作主语稍显冗长;将“selection”和“use”转译为动词,更加清楚流畅。

[19] 原译“运输及运输时间”令人费解:“运输”包括“运输时间”,查看原文件得知,transport 板块更多在规定其方式,因此改译为“运输方式及运输时间”。

[20] 英语中的“unless otherwise stipulated/ agreed upon/ mentioned”等句式多置于句首,汉语法律文件的“除……外”多置于句末。

[21] 根据 Law Insider 网站(https://www.lawinsider.com/dictionary/manufacturers-certificate)的解释,“Manufacturer's Certificate”指“any Manufacturer's statement of origin, certificate of origin or any other document evidencing the ownership or transfer of ownership of a New Motor Vehicle from a Manufacturer to a Borrower”。因此,“Manufacturer's Certificate”是一纸文件,用于证明某事,而不是“证书”,因此原译的“证书”应改为“证明(文件)”。“厂商”和“制造商”在这里没有明显区别。另外,参考译文将句中的对象提前,起到强调作用,同时也使上下文更连贯。

[22] (1)原译中“对……保留权利”“对……进行测试”有较重的“翻译腔”色彩,参考译文将“reserve the right”灵活处理为“有权”。(2)工程领域中的“Employer”在 1987 年前译为“业主”,之后译为“雇主”。(3)“lot”在工程英语中多指“批次”,所以这里的“batch”和“lot”实则同义。(4)“test”对应的中文有“试验”“实验”“检测”“检验”等,结合全文,此处应为“检测”或“检验”的意思。

[23] (1)原译严格按照英文语序翻译稍显冗杂,根据汉语特点可以省译个别动词和介词。(2)这句话和上一句话的后半部分有明显的逻辑关系,因此参考译文将两者用逗号相连以并列句形式处理译文。

[24] 法律文件中的“shall not”一般译为“不得”。“...shall not be permitted”完整的意思是“不得允许”,考虑到主语涉及的人员相对复杂,在无法确认人员范围的情况下,参考译文灵活处理为“不得”,乃是权宜之举。

[25] 原文中“sulphate content”(硫酸盐含量)可能指水泥中的硫酸盐,也可能指施工现场岩土中硫酸盐的含量,这时“Geotechnical investigation”(岩土工程勘查)字眼至关重要,这项勘查肯定不是勘查水泥。另外,根据土木工程相关知识,抗硫酸盐水泥主要用于受

硫酸盐侵蚀的土地。据此,增译"工程岩土"可使译文更清晰易懂。

[26] 这句话非常复杂,翻译分为四步。第一步,将信息拆分如下:①水泥的 curing time 将在 execution of the work(动工)时决定。②某样东西是按照合同约定(under the contract)的。③curing time 要根据知名实验室的(水泥检验合格)证明(based on the certificate from a reputed laboratory)。④报告由承包商获得并提交。其中,"under the contract"修饰的是动工时间。第二步,解决关键词的意思。①"curing time",有可能译为"凝固时间""固化时间""养护时间"等。其中,"凝固时间"或"固化时间"虽然会受温度、气候等影响,但一般比较固定,因此不需要人去决定。据此可以推测,"curing time"可能是"养护时间"。然后,经查询水泥/混凝土(里面有水泥)的确需要养护。综上,可以断定,"curing time"在该语境中应译为"养护期"或者"养护时间"。②"reputed"一般解释为"according to what some people say, but not definitely",类似"据说""道听途说",而不是"trustworthy"或"respected"(信誉良好)的意思。但是,在英语搜索引擎中输入该词,得到其还有"well known"的意思。本文本中的用法可能与印度英语有关。第三步,试译。尝试翻译为"水泥的养护时间将在承包商获得并提交知名实验室的证明后,在合同约定的开工时间决定。"可以理解,但仍有瑕疵。第四步,改译。"提交知名实验室的证明"中应增译"出具",即"提交知名实验室出具的证明"。"certificate"应增译出完整的意思,即"水泥检验合格证明"。然后根据汉语习惯,按时间先后顺序和逻辑顺序,增加小标题,重新加以编排。

[27] 根据前面的讲解,增译"印标",同时"IS 383 (1970)"应与官方写法一致。

[28] 原译"水是新鲜的、干净的"有较重的"翻译腔"色彩,这里相当于"clean, fresh water",即"干净的淡水"。"油、酸和碱、有机材料以及其他有害物质"在语法上并没有错误。但此处对搅拌用水来说,"油、酸、碱、有机材料"都是有害物质,所以可以简化成"油、酸、碱、有机材料等有害物质"。

[29] 此句话的完整表达应为"Reinforcement shall conform to IS: 432-1966 for M. S bars and hard drawn steel wires and reinforcement shall conform to IS: 1138-1966 and IS: 1786-1966 for deformed and cold twisted bars respectively",简化为"bars and hard drawn steel wires shall conform to IS: 432-1966 and deformed and cold twisted bars (shall conform) to IS: 1138-1966 and IS: 1786-1966"。"bars"很容易直接翻译成"棒"或者"棒条",但是土木工程行业里一般不说"低碳钢筋棒(条)",只说"低碳钢筋",同理,行业里一般只说"变形钢筋"和"冷轧扭钢筋"。

[30] 原译"图纸和说明书"不够具体,通过检索"施工 specifications",可以查到"施工细则"这一用法。为使意思更清楚,参考译文增译了两个"施工"。(3)"place"可视情况译为"堆放"(钢筋)或"安装"(钢筋),这里有"dimensions"(尺寸)作为参考,一般不会规定按某个形状和尺寸堆放,而是按这种方式安装。

[31] 原译中的“地基的指示性形状”让人无法理解；这里的“indicative”应当是由“as indicated”变化而来，因此应译作“所需地基基础的形状”。

[32] “Price Schedule”两个单词首字母大写，说明其为专有名词，且这是一个表格，汉译时应在括号内标注对应英文。“Price Schedule”整体的意思是《价格附表》，但其指代不够具体，因此根据上下文将其显化为《投标报价表》。“tower”在该语境中指的是“高楼”或“大厦”，不宜译为“塔”或“塔楼”。

[33] (1)“provide”的宾语为三种“quantities”（工程量），对应三种单价。原文为避免重复，用复数“rates”表示三种价格，汉译时需使用三个“单价”字眼，即“开挖单价、混凝土单价及钢筋单价”。在这一点上，原译的“挖掘、混凝土和钢筋的单价”有明显的“翻译腔”痕迹。(2)“as classified in clause 2.2 of this section”对“type of foundation”起补充说明作用，翻译时应置于“各类基础”后，这一点原译处理也有失误。此外，可输入“法律 clause section”等查询法律章、节、条、款、目等的对应说法。正式文件中，一般应使用汉字翻译阿拉伯数字。相关要求请参阅《中华人民共和国国家标准出版物数字用法》(GB/T 15835—2011)。(3)“guaranteed foundation quantities”较难翻译。其一，“quantities”后面括号中的“Excavation volume”是“开挖量”，“Concrete volumes”是“混凝土体积”或者“混凝土方量”。“开挖量”和“混凝土体积”都属于工程量，因此“quantities”应译为“工程量”，而不是原译中的“数量”。其二，“guaranteed”的意思是“能够起保障作用的”，这个语境下，译为“保障”比较简洁。其三，本句参考译文前半部分已有“基础”这一信息，因此“guaranteed foundation”中的“foundation”可以省译。综上，将“guaranteed foundation quantities”译为“保障工程量”最为简洁、准确。

[34] 英文原文简化后的结构为 composite price must comply unit rate and quantities，直译为“综合价格必须符合单价和工程量”，这与事实不符。查询得知，综合价格、单价、工程量之间是乘积和乘数的关系，因此参考译文增译了“乘积”一词。

[35] 原文“measurement shall be in cubic meters”中“measurement”指“units of measurement”，因此增译了“单位”一词。但是，这句话前半部分主语是“计量单位”，跟后面的“计算”逻辑矛盾。参考译文进行了词性转换，灵活处理为“测量混凝土体积时，应当用‘立方米’计量，计算时保留两位小数”。

[36] 地基加固一般不需要用钢材，基础加固才需要用钢材。因此，此处“foundation”仍然译为“基础”。

[37] “修改到小数点后 3 位”不符合汉语常见表达，更常见的是“保留到小数点后第 3 位”或“保留小数点后 3 位数字”。“make allowance for”的意思为“考虑到……”，此处“allowance”并非“津贴”的意思。

[38] (1)结合全文背景知识来看，“wire”并非“电线”，而是“细线”，用于绑扎的线(wire required for binding)即“绑扎线”。“wire required for binding”限定后面所有并列成分，

因此需要将“绑扎线”一词重复译出。(2)英语中有大量的“be regarded as”“be deemed as”“be taken as”等结构,汉译时多数时候不必出现“被”字。(3)原文的as(原因)部分比较长,根据状语从句的特点可调整语序,置于句末;中文通常先说原因,且“as”的程度不如“because”深,汉译时可以省略“as”这个连接词。省略后,形散意不散,更符合汉语重意合的特点。

练习题

一、英译汉

Surveying is one of the oldest activities of the civil engineering and remains a primary component of civil engineering. It is also one field that continues to undergo phenomenal changes due to technological developments in digital imaging and satellite positioning. These modern surveying tools are not only revolutionizing regular surveying engineering tasks but are also impacting a myriad of applications in a variety of fields.

Modern surveying engineering encompasses several speciality areas, each of which requires substantial knowledge and training in order to attain proper expertise. The most primary area perhaps is plane surveying because it is so widely applied in engineering and surveying practice. In plane surveying, we consider the fundamentals of measuring distance, angle, direction, and elevation. These measured quantities are then used to determine position, slope, area, and volume-the basic parameters of civil engineering design and construction. Plane surveying is the measurement of the earth's surface as though it were a flat surface without a curvature. Within areas of about 20 kilometers square-meaning a square, each side of which is 20 kilometers long-the effects of the earth's curvature are negligible relative to the positional accuracy. For larger areas, however, a geodetic survey, which takes into account the curvature of the earth, must be made.

Geodesy, or higher surveying, is an extensive discipline dealing with mathematical and physical aspects of modeling the size and shape of the earth, and its gravity field. Since the launch of earth-orbiting satellites, geodesy has become a truly three-dimensional science. Terrestrial and space geodetic measurement techniques, and particularly the technique of satellite surveying using the Global Positioning System (GPS), are applied in geodetic surveying. GPS surveying has not only revolutionized the art of navigation but has also brought about an efficient positioning technique for a variety of users, prominent among them the engineering community. GPS has had a profound impact on the fundamental problems of determining relative and absolute positions on the earth, including improvements in speed, timeliness, and accuracy. It is safe to say that any geometry-based data collection scheme profits to some degree from the full constellation of 24 GPS satellites. In addition to the obvious applications in geodesy, plane surveying, and photogrammetry, the use of GPS is applied in civil engineering areas such as transportation (truck and emergency vehicle monitoring) and structures (monitoring of deformation of structures such as water dams).

Photogrammetry and remote sensing encompass all activities involved in deriving qualitative and quantitative information about objects and environments from their images. Such imagery may be acquired at close range, from aircraft, or from satellites. In addition to large-, medium-, and small-scale mapping, many other applications such as resource management and environmental assessment and monitoring rely on imageries of various types. Large-scale mapping (including the capture of data on infrastructure) remains the primary civil engineering application of photogrammetry.

二、汉译英

所有支承在地基上的结构物,包括建筑物、桥梁、土堤以及土坝、土石坝、混凝土坝,都是由两部分组成的。它们是上部结构,或者说上面部分,和介于上部结构与支承地基之间的下部结构构件。就土堤和各种坝而言,在上、下部结构之间通常没有一条明显的划分线。"基础"这个词可以定义为下部结构及其邻近的土和/或岩石,这些土和/或岩石受到下部结构构件与荷载的共同影响。

基础工程师是能够根据经验和所受到的培训对涉及这一部分工程系统的设计问题提出解决方案的人员。就此来看,基础工程可以定义为运用土力学与结构力学原理和工程判断来解决交界面问题的科学和技术。基础工程师直接关心的是影响上部结构向地基传递荷载的结构构件,使地基稳定性及估算的变形量都在允许范围内。由于下部结构构件的设计几何尺寸和位置通常会对地基有影响,因此基础工程师必须具有足够的结构设计方面的知识。从最小的住宅房屋到最高的高层建筑以及桥梁等结构物的基础都是用来传递上部结构荷载的。这种从柱型构件传下的荷载的应力大小,钢柱大约在 140 兆帕,混凝土柱大约为 10 兆帕。而土的承载能力则很少超过 500 千帕,比较常见的是 200 ~ 250 千帕。

任何合理的结构物,只要有足够的经费,通常可以建造得很安全。可惜的是,在实际工作中即使存在着这种情况,也是非常少有的,基础工程师总需要在与理想情况相差很多的困境下做出决定。此外,即使可以掩埋错误,但无法掩盖错误造成的后果,并且能够较快地显示出来——而且可能会在某些法规所规定的期限到期之前出现失效。已经有过这样事例的报道,基础的缺陷(如墙体开裂)是在施工结束几年之后才显示出来的,但是,在有的事例中,基础的缺陷是在上部结构施工过程中或者在施工结束后马上显现出来的。

设计人员总是面临如何进行既安全又经济的设计,同时又要与必然存在的工地天然土壤的不匀问题作斗争。现在又因为土地稀少,需要利用那些曾经用作垃圾填埋场,甚至危险废料处理场的土地,致使这个问题变得更加复杂。还有另一个复杂因素是施工作业会改变土壤性质,使其相对它在基础的初步分析、设计时发生显著的变化。这些因素都会使基础设计难以量化,因此两个设计单位可能会提出完全不同的设计,但其使用效果可能会同样令人满意。

第十六章
材料工程类

【汉语原文】

低塑性抛光技术综述(节选)

引言

80%以上的结构部件失效是由材料表面萌生的疲劳失效引起的,尤其是航空发动机零部件占更大比例[1]。因此,优化材料表面完整性对提高金属材料构件的疲劳性能具有重要意义[2]。低塑性抛光技术可以有效地提高金属材料的表面疲劳性能,已广泛应用于诸多工程领域[3]。

通常,早期疲劳失效发生在材料的表面或亚表面,然后逐渐扩展到材料的内部,最终导致部件的完全失效[4]。因此,大量研究表明,材料表面对材料的使役性能有重大影响[5]。正因如此,大量的研究者致力于材料表面完整性的研究,包括表面状态和性能[6]。

鉴于航空发动机的特殊使用环境,材料需要可靠性高、寿命长、成本低[7]。因此,材料的表面性能非常重要[8]。此外,目前航空工业中使用的许多金属材料,如钛合金、铝合金、不锈钢,以及镍基高温合金,在越来越复杂的使用条件下表现出较差的疲劳抗性[9]。因此,许多研究人员通过表面强化方法来提高材料的表面强度和表面完整性,延迟或抑制疲劳裂纹的萌生和扩展[10]。

低塑性抛光技术是典型的表面机械强化技术[11]。通过施加机械载荷使材料表面发生塑性变形,使表面完整性得到明显改善,从而提高使役性能[12]。此外,表面机械强化技术无须外部热源或热处理过程的相变[13],这些技术在航空、汽车、机械、焊接等行业得到了广泛应用,尤其是在蒸汽轮机和航空发动机叶片领域[14]。到目前为止,已经开发了很多表面机械强化方法,如超声纳米表面改性(UNSM)、喷丸强化(SP)、超声喷丸(UP)、超声表面滚压(USRP)、表面机械研磨(SMAT)、表面机械滚压(SMRT)、低塑性抛光(LPB)、水喷丸和空化喷丸(WJP, CP)、激光冲击强化(LSP)[15]。

在这些表面机械强化技术中,LPB方法相比其他方法具有一定的优势,包括与车床同步工作的简单机械结构、低成本、高效率、低环境污染和高可控性[16]。LPB已经成功应用于燃气轮机压缩机叶片的表面强化,显著提高其抗疲劳性能和抗异物损伤性能[17]。此外,大量研究已证明,LPB技术可以强化多种类型的金属材料[18]。

本综述研究综合分析LPB工艺对不同金属材料的表面完整性和使役性能的影响,以及

LPB 与其他表面机械强化方法的对比[19]。此外还总结了 LPB 在将来的研究和发展所面临的挑战和相应的建议,这些将会促进表面机械强化技术的发展和应用[20]。

【文本分析】

原文本节选自某低塑性抛光技术综述论文(*A Review of Low-Plasticity Burnishing and Its Applications*)的引言部分,为便于编排,本教材编写团队删除了部分内容及文内引用。中文文章由作者撰写,原译由其他译员提供,参考译文由本教材编写团队成员提供。改译后,作者团队做了极小程度的调整。文章最终发表于全球知名出版社 Wiley Online Library 旗下的学术期刊 *Advanced Engineering Materials*,全文获取地址为 https://doi.org/10.1002/adem.202200365。

原文属于学术类科技类文体,兼具学术特点和科技感。学术文本和科技文本均属于信息类文本(informative texts),该类文本的理想翻译风格为"中立、客观",翻译方法为等效翻译,最大翻译单位为句子,最小翻译单位为搭配。因此,参考译文在语篇衔接上做的调整较小,更多是在句子内部做出调整,同时修改部分搭配、累赘表述等。另外,学术文章的风格应尽量满足期刊出版社指南的要求,翻译前应尽可能阅读相关写作风格指南。

原文的目标受众是科研人员,因此译文中专业术语应当准确。节选部分的中文表述比较清楚,大大降低了翻译难度。值得一提的是,本章的参考译文为期刊已接受的版本,但由于改译人员时间极为有限,部分译文仍有提升的空间,这在译文详解部分将提及。

【原译】

A Review of Low-Plasticity Burnishing and Its Applications(excerpt)

Introduction

More than 80% of structure component failure is caused by the fatigue failure that initiates from material surface, especially for the aero-engine parts account for a larger proportion[1]. Therefore, optimizing the surface integrity of material to improve the fatigue properties of metallic material components is of significance[2]. Low plastic polishing (LPB) technology can effectively improve the surface fatigue properties of metallic materials, and it has been widely used in various engineering field[3].

Generally, the early fatigue failure occurs on the surface or subsurface of the material, and then gradually expands to the interior of the material, and eventually leads to the complete failure of the components[4]. Therefore, a large number of studies have demonstrated that the surface of materials has significant impact on the use properties of materials[5]. Because of this, a large number of researchers are devoted to the study of the integrity of material surface, including surface state and properties[6].

In view of the special use environment of aero-engine, the high safety, long life and low cost are required for materials[7]. Therefore, the surface properties of materials are very important[8].

In addition, many metallic materials, such as titanium alloys, aluminum alloys and stainless steels, as well as nickel-based superalloys, currently used in aviation industry show poorer fatigue resistance under more and more complex use conditions[9]. Therefore, numerous investigators try to improve the surface strength and surface integrity of materials through surface strengthening method, and further to delay or inhibit the initiation and expansion of fatigue crackles[10].

LPB technology is the typical surface mechanical strengthening technology[11]. Plastic deformation occurred on the material surface by the applying mechanical load, resulting in a significant improvement of surface integrity, which could improve the service performance[12]. Besides, the advantages of without external heat source and heat phase transformation promote that[13] various surface mechanical strengthening technologies are widely used in aviation, machinery, automobile, welding and other industries, especially in the field of steam turbine and aviation engine blades[14]. So far, many surface mechanical strengthening methods have been successfully developed, such as Ultrasonic Nanocrystal Surface Modification (UNSM), Shot Peening (SP), Ultrasonic Peening (UP), Ultrasonic Surface Rolling Processing (USRP), Surface Mechanical Attrition Treatment (SMAT), Low Plastic Polishing (LPB), Water Jet Peening and Cavitation Peening (WJP, CP), Laser Shock Peening (LSP)[15].

Among these surface mechanical strengthening technologies, the LPB method has some advantages, including simple mechanical structure synchronously working with lathe, low cost and high efficiency, low environmental pollution, and high controllability, when compared with other methods[16]. LPB has been successfully used in surface strengthening of gas turbine compressor blades, resulting in a significant improvement of the resistance of fatigue and foreign body injury[17]. In addition, a larger number of types of metallic materials can be strengthened by LPB, which has been proved by many researches[18].

In this review, the influence of LPB process on surface integrity and service performance of different metallic materials, as well as the comparison among LPB and others surface mechanical strengthening methods, were compressively analyzed[19]. Besides, some challenges faced and suggestions on the directions of future research and development for LPB were given, which has important significance for the development and application of surface mechanical strengthening technology[20].

【参考译文】

A Review of Low-Plasticity Burnishing and Its Applications(excerpt)

Introduction

Fatigue failure that is initiated on material surfaces causes more than 80% of structure component failures, an even larger proportion for aeroengine parts[1]. It is, therefore, crucial to improve the fatigue performance of metallic components by optimizing their surface integrity[2]. Since fatigue performance can be effectively improved via low-plasticity burnishing (LPB) technology, LPB has been widely used in various engineering fields[3].

Generally, the early fatigue failure occurs on the surface or subsurface, then gradually expands to the interior, and eventually leads to complete failure of the components[4]. Hence, multiple studies have demonstrated that the surface of materials has a significant impact on their service performance[5]. As a result, numerous investigators have long been devoted to the material surface integrity, including surface state and performance[6].

Considering the special service environment of aeroengine, materials need to feature high reliability, long life, and low cost[7]. Therefore, the surface performance of materials is critical[8]. In addition, many metallic materials currently used in the aviation industry, such as titanium alloys, aluminum alloys, stainless steels, and nickel-based super alloys, show poor fatigue resistance under increasingly complex service conditions[9]. Numerous investigators have therefore used surface strengthening methods to improve the material surface strength and integrity, thereby delaying or inhibiting the initiation and propagation of fatigue cracks[10].

LPB technology is one of the typical surface mechanical strengthening technologies[11]. The mechanical load applied on the material surface induces plastic deformation, a result that can significantly improve surface integrity and service performance[12]. In addition, since surface mechanical strengthening technologies do not need external heat sources or phase transformation by heat treatment[13], these technologies have been widely used in aviation, automobile, machinery, and welding industries, especially in steam turbines and aircraft engine blades[14]. So far, scientists have developed many surface mechanical strengthening methods, such as ultrasonic nanocrystal surface modification (UNSM), shot peening (SP), ultrasonic peening (UP), ultrasonic surface rolling processing (USRP), surface mechanical attrition treatment (SMAT), surface mechanical rolling treatment (SMRT), LPB, water jet peening and cavitation peening (WJP, CP), and laser shock peening (LSP)[15].

Among the abovementioned technologies, the LPB method has certain advantages over others, including a simple mechanical structure that works synchronously with the lathe, low cost, high efficiency, low environmental pollution, and high controllability[16]. This method has been used in the surface strengthening of gas turbine compressor blades, significantly improving their resistance to fatigue and foreign object damage (FOD)[17]. In addition, the method can strengthen substantial types of metallic materials, as has been proved by many researchers[18].

In this review, we comprehensively analyze the influence of the LPB process on surface integrity and service performance of different metallic materials and compare LPB and other surface mechanical strengthening methods[19]. In addition, we summarize the challenges to, and corresponding suggestions on, LPB's future research and development, all of which may facilitate the development and application of surface mechanical strengthening technology[20].

【译文详解】

[1] (1)原译为"More than 80% of structure component failure..., especially for the aero-engine parts account for a larger proportion",其中"80%"是一个比例,跟句末的"比例"(proportion)这一密切相关的信息间隔太远,导致可读性较差。由于原文意思等同于"由材料表面萌生的疲劳能引起 80% 的结构部件失效",因此参考译文将"80%"后移,处理为"...80% of structure component failures, an even larger proportion for..."。(2)原译"especially for"后接句子属于语法错误;语法正确的版本应为"especially for the aeroengine parts which/that account for...",但比较冗长。参考译文省略了原文中"尤其是"这一并无实际意义的部分,改用"an even larger proportion for aeroengine parts"这一简单结构。(3)疲劳失效(fatigue failure)是由其他原因引起的,因此参考译文将原译的主动语态改为被动语态。

[2] (1)原译"optimizing...to improve...properties...is of significance"的句型头重脚轻且具有中式英语色彩。汉语以述评(Topic-Comment)结构为主要句法特点,即先说主题再评论或补充说明。原句的"优化材料表面完整性"为主题(topic),"具有重要意义"为评述。但是,英语中简短的评述部分多放在句首。因此参考译文采用"it is...to improve...performance...by optimizing"句式。(2)"therefore"一词放在句首的情况很少。该词一般放在①连词 and 后(如 Their car was bigger and therefore more comfortable),②放在 be 动词、情态动词或 will、would、has、have 等时态屈折词后(We will therefore do...),③放在主语后实义动词前(如 I therefore propose...)。特殊情况,将在本章[8]中讲解。(3)原译"of significance"不够简洁,参考译文用"crucial"更直接。(4)"优化材料表面完整性对提高金属材料构件"中的两个"材料"是同一物体,英译时应避免重复,因此参考译文使用物主代词"their"替换"of the metallic components"。(5)原译中的"properties"为特性,也就是特殊的属性,"性能"的英语一般为"performance",强调其表现的能力。

[3] (1)英语写作的信息流动(information flow)的方式是:old-new-old-new。原句的"金属材料的表面疲劳性能"是上一句刚提到的信息,为旧信息,因此英译时置于句首更为流畅、更地道。当然,旧信息提前,主语和宾语顺序颠倒,势必要用到被动语态。(2)这句话和上一句有逻辑关系,汉语重意合,没有逻辑连接词并不影响理解,但英译时应该将逻辑显化出来。因此,参考译文增译了"since"一词。(3)"低塑性抛光"的英语为"low plasticity burnishing",而非"low plastic polishing"。另外,专业术语首次出现时一般需用

全称，然后在括号内写出简称，以便下文使用。

[4] (1)原译在一句话里出现了两个“of the material”，一般保留一个，以避免重复。参考译文根据上下文语篇逻辑，将两个“of the material”全部省略，既没有歧义又十分简洁。(2)参考译文没有删除“the components”中的“the”，这里没有特指，删除更合适。(3)“expand”和“extend”很多时候可以互用，但“extend”更强调“length”或在一个方向上的延伸，“expand”强调“area”或各个方向的延展。参考译文在未求得作者证实的情况下，未轻易改动。此外，从下文的查询来看，这里还可以用“propagate”一词。

[5] (1)“a large number of”不够简洁且句式比较普通；参考译文更简洁，词语使用稍显高级。(2)原译的“has significant impact on”含语法错误：“impact”作不可数名词时，意为“冲击力”；作“影响”意时，通常用单数，且需要加冠词。(3)原译将“使役性能”译为“use properties”属于术语误译。本章[2]中已经提及“性能”的正确英译为“performance”，而“使役”的完整意思为“使用服役”(https://thinktank.sciencereading.cn/booklib/v/subLibPreview/122/245/993013.html)。“使用”的英语为“use”，“服役”的英语为“service”，因此不妨在英语搜索引擎中输入“use and service performance of material”加以验证，得出词频为0。而输入“service performance * material”出现了一些词条，并且在Metals & Minerals - CWA International官网(https://www.cwa.international/metals_and_minerals.php)有相应说法。

[6] (1)原译“...researchers have been devoted to the study of the integrity of”不够简洁。“researchers”已经暗含“study”或“research”“研究”之意了，因此参考译文省略了“the study of”。(2)原译将“材料表面完整性”译为“integrity of material surface”，但这一表达不太符合行业习惯用法，英语搜索引擎查到的词条也寥寥无几。在英语搜索引擎中输入另外两个可能的版本“material surface integrity”和“material's surface integrity”，结果“material surface integrity”词条较多且多为专业领域用词，因此参考译文选择了这一表述。

[7] (1)“鉴于”的英语表达有“in view of”“given”“considering”等。第一个偏长，不够简明；最初改译为“given”；“considering”用在这里是垂悬结构(dangling modifier)，但也是出于这个考虑没有用“considering”，不过这个跟“judging from”等表达一样属于错误但可以接受的垂悬结构。(2)原译“In view of the special use environment of aero-engine, the high safety, long life and low cost are required for materials.”介词结构和主语稍长，使用被动语态显得有些头重脚轻，可读性较差，且不符合该期刊出版社建议多使用主动语态的要求。

[8] 此处“therefore”的摆放位置为特殊情况。要是将“therefore”要是放在主语后，本句将变成“The surface performance of materials is therefore critical”。这样的摆放方法不利于“therefore”连接前一句句末的“high reliability, long life, and low cost”三种特性与和后面

的“performance”,有损信息流动(information flow)。

[9] (1)原译“In addition, many metallic materials, such as…, currently used…”中“currently used”部分离被修饰成分“many metallic materials”太远,可读性较差。因此参考译文将其挪至“materials”后,改为“In addition, many metallic materials currently used…, such as…”。(2)原译“aviation industry”缺少定冠词。专门的行业前均需要加“the”。类似的还有“文化产业”(the cultural industry)。(3)原文“表现出较差的疲劳抗性”中的“较差”并没有真正跟其他类别比较,原译为“poorer”属未理解原文的深层次意义。因此,参考译文根据实际语境,改译为“poor fatigue resistance”。(4)参考译文中的“increasingly”比原译中的“more and more”更为简洁,且在学术语境中使用频率更高。

[10] 本句较为复杂,原译问题较多。(1)原译“numerous investigators try to”时态错误。学术写作中表达众多学者研究某课题的现象时,一般用完成时态。(2)原译将“通过”部分译为介词结构,信息主次不清。从上下文看“完整性”并不属于新信息或重要信息,重点信息是“用技术”。基于此,参考译文将“improve…through…method”的结构改为“use method to improve…”的结构。(3)原译“improve the surface strength and surface integrity of materials through surface strengthening method”中有三个“surface”,较为冗余。参考译文省译了一个“表面”,用“surface strength and integrity”更为简洁。另外,最后一个“surface strengthening”中的“surface”省略后可能有歧义,因此参考译文将其保留。(4)原文“提高……表面完整性,延迟或抑制……”是典型的意合结构,逗号前后的内容存在因果关系或递进关系。原译将其处理为“and further to”和前半句的“to improve”并列,逻辑错误。参考译文用“improve, thereby delaying…”逻辑更清晰。(5)“疲劳裂纹”译为“fatigue crackle”术语错误:“crackle”的意思(爆裂声)跟裂纹一词毫不相干。该词可译为“fatigue cracking”(指发生裂纹这种现象)或“fatigue cracks”(指具体的裂纹)。(6)“萌生和扩展”可以在网页里输入“疲劳裂纹的萌生与扩展 英语”,得到的结果除了“sprouting and extending”外,基本都是“initiation and propagation”,将这个术语放到英语搜索引擎里验证即可。

[11] 前文括号中备注了简称,这里用“LPB”这个简称即可。“典型的”说明并不是唯一的,原译用“the”有误,应该用“a”或用“one of the”。

[12] (1)原译“Plastic deformation occurred on the material surface on the material surface by the applying mechanical load”跟前文时态不一致。另外,“by the applying”应该为“by applying”或“by the application of”才对;“by applying”比“by the application of”更简洁有力。(2)原文“通过……使”句式杂糅,正确的理解应该是“施加机械负荷(能)使材料表面发生塑性变形”或“施加在材料表面的机械负荷(能)使(材料表面)发生塑性变形”。参考译文采用了第二种处理方式。(3)原译“resulting in a significant improvement of”十分迂回,改译“significantly improve”更简洁有力。(4)参考译文中的“a result”是

对前一句话的总结,功能相当于“which”。值得注意的是,“which”引导的非限制性定语从句既可以修饰其前面的名词,也可以修饰整个句子,常常会产生歧义。有歧义时,想要修饰整个句子,一般先用“a/an + 名词”总结前面的句子,再用 that/which 引导从句。(5)参考译文中“service performance”前加上“therefore”一词逻辑会更清楚。

[13] (1)原文逗号前后存在明显的逻辑关系,原译也意识到了这一点,不过处理得过于复杂且语法错误明显。“advantages promote”这种搭配也十分罕见。原文其实是简单的因果关系,不必处理得过于迂回。(2)“热处理过程的相变”意思是“相变是在热处理过程中发生的”,译为“phase transformation(that results)from heat treatment”更好。(3)“相变”指固、液、气三相之间的转变,应译为“phase transition”或“phase change”。原译和参考译文均用了“transformation”(形态变化),实则欠妥。但是,原文本写作时参考的文献使用了“phase transformation”,为尊重原作者,参考译文保留了原译的说法。

[14] (1)“used in aviation, machinery, automobile, welding and other industries”中的“and other industries”为汉语“等”字翻译而来,句式比较中式。如果此处“等”确实指未列举完毕,英语更常用的是“in industries such as A, B, C, and D”或者“in such industries as A, B, C, and D”。另外,根据《现代汉语词典》(第7版)的解释,“等”有时候只是用于列举后煞尾,并无实际语义功能,因此参考译文省译了“and other industries”。

[15] (1)在不知道动作发出者为谁的情况下,原译使用被动语态没有问题,但参考译文增译了“scientists”,使用主动语态,更符合大多数出版社的写作风格要求。(2)列举中“and”一词前是否需要逗号(serial comma,也称 series comma, Oxford comma, or Harvard comma),应根据期刊风格指南要求处理。英式英语一般不加 serial comma,但有歧义时一定要加,如“The jumper is available in green, yellow, and black and white.”(这里“black and white”是一个整体)。美式英语一般建议甚至强制要求加 serial comma,但也有例外(详见 https://en.wikipedia.org/wiki/Serial_comma)。(3)原译中“LPB has been successfully used”有些中式表达,“developed”用完成时态,已经暗含了成功的意思(没有人会说“失败地 developed”)。因此参考译文省译了“成功”一词。(4)本句中的众多专有名词都比较常见,在 SCIdict 学术词典和术语在线网页中可以查到。

[16] (1)“在这些表面机械强化技术中”原译为“Among these surface mechanical strengthening technologies”,从语篇角度看来十分冗余。本句前一段一直在讲“mechanical strengthening technologies”,因此参考译文根据语篇衔接省略了“表面机械强化”,译为“Among the abovementioned technologies”。当然,译为“Among these technologies”亦可。(2)“与……相对,更……”结构可译作“compared with ABC, XYZ...”结构,但原译“the LPB method has some advantages, ..., compared with ...”中“advantages”与其相关信息“when compared with other methods”离得太远,影响可读性。原文“LPB 方法相比其他方法具有一定的优势”可理解为“LPB 比其他方法更有优

势”,因此参考译文灵活处理为“LPB has certain advantages over others”,更简洁、地道。(3)原文中“简单机械结构”从技术层面看是可数名词,因此参考译文增译了不定冠词“a”,原文中的“车床”(lathe)虽然没有明确是特指,但可以根据语境推测出来,因此参考译文增译了“the”(the lathe)。(4)原译“including simple mechanical structure synchronously working with lathe”中有两个非并列的V-ing,风格不够简明。参考译文将后置定语“working...”改为“that works...”,虽然多一个词,但可读性较强。

[17] (1)原译上一句主语为“the LPB method”,本句仍将“the LPB method”用作主语,显得有些重复。参考译文用“this method”不仅指向性更明确,而且相对简洁。另外,“strengthening”是动名词,后接“of”时,其前应增译定冠词“the”(即the surface strengthening of)。(2)原译“the resistance”指向性不明确,改译为“their resistance”。(3)“提高抗异物损伤性能”是指抵抗异物损伤的性能。首先考虑“improve the performance of resisting...”,但在该语境中“resistance”本身就是一种性能。因此参考译文省译了“the performance of”,译为“improve their resistance”。(4)原译“异物损伤”,术语错误。网页搜索“异物损伤(空格)英语”,结果出现“foreign body injury”“foreign object injury”“foreign object damage”“foreign body damage”四种译法。首先排除医学用语的“injury”,再用英语搜索引擎加以验证。“foreign object damage”在维基百科中有专门的词条,且备注“FOD”简写字样。因此参考译文选用了“foreign body damage (FOD)”。

[18] (1)新旧信息的排放顺序请参考本章[3](1)的讲解。此处,LPB是旧信息,应置于句首。这样一来,原译的被动语态自然变成了更为简明的主动语态。(2)原译的“which”略有歧义,参考译文使用的“as”更清晰明了。另外,原译中“many researches”不太符合英语表达习惯;“research”更多是作单数,作复数时,更常用的表达是“many pieces of research”,不过在此处稍显冗余;参考译文根据语篇逻辑灵活处理为“researchers”。(3)参考译文中“as has been proved by many researchers”可进一步改为更简明的主动语态版本“as many researchers have proved”,且这部分可以放在主语后作插入语。(4)原译使用了好几个“a large number of”这种冗长的表达,参考译文使用了“a multiple of”等高级表达或更简洁的“many”。

[19] (1)原译明显头重脚轻。原文使用的“doer + action”结构,英语也有一样的结构,没有必要为追求所谓的学术感而使用头重脚轻的被动语态。原译中的“the influence of... on...of”结构中名词和介词过多,可读性较差。参考译文使用主动语态,将原文的名词转换为动词,译为“we comprehensively analyze how the LPB process influence the surface...”更简洁有力,更符合期刊风格要求。(2)原译“comparison was analyzed”动作发出者模糊,且语言不够简洁,不符合期刊风格要求。此处是作者团队在“compare”某些技术,因此参考译文使用“we compare...”这一简洁明了的表述。

[20] (1)原译“challenges faced”有语法错误,应该是“challenges something/somebody faces”

"challenges something/somebody were faced with"或"challenges facing something/somebody"。另外,参考译文将动词转换为介词"to"更加简洁。(2)原译将"这些"译为"which",指代不清。从语法层面看,"which"一词或指整个句子,或指离"which"最近的名词"future research and development"或"the directions of future research and development",但原文的"这些"指的是"挑战和建议",跟原译语法上所指的内容完全不一样。从文章后面来看,作者团队总结的挑战和提出的意见不止两条,因此参考译文使用了 all of which(复数概念),指向性更明确。(3)原译中"facilitate the development and application of"比较冗余,参考译文将其简化为"help to develop and apply"。(4)原文中的"技术"不止一种,因此原文和已出版译本中的"technology"改为"technologies"更好。(5)"表面机械强化"的原译为"surface mechanical strengthening",不太符合英语表达习惯(英语几乎不用名词+形容词+名词/动名词的结构)。对此本教材编写团队曾提出质疑,但作者团队表示,原译可以接受。"表面机械强化"的意思是在材料的表面使用机械强化技术去强化该表面,通过在术语在线网站中检索可以确定"表面机械强化"确应译为"mechanical surface strengthening"。

练习题

一、英译汉

Powder metallurgy which is also called powder forming, is a process of making components from metallic powders. Initially, it was used to replace castings for metals which were difficult to melt because of high melting point. The development of technique made it possible to produce a product economically, and today it occupies an important place in the field of metal processing. Powder metallurgy manufacturing technology consists of three steps: mixing metal or alloy powders, compacting those powders in a die at a room temperature and then sintering or heating the shape in a controlled atmosphere furnace to bond the particles together.

Mixing of Powders

Mixing of the powders is essential for uniformity of the product. Lubricants are added to the blending of powders before the mixing. The function of lubricant is to minimize wear, to reduce friction. Different powders in correct proportions are thoroughly mixed in a ball mill.

Compacting

The metal powders are placed in a die cavity and compressed to form a component shaped to the contour of the die. The pressure used for producing the green compact of the component ranges from 80 MPa to 1400 MPa, depending upon the material and the characteristics of the powders used. Mechanical presses are used for compacting objects at low pressure, while hydraulic presses are used for compacting objects at high pressure.

Sintering

Sintering involves the heating of a green compact at a high temperature in a controlled atmosphere. Sintering increases the bond between metal particles and therefore strengthens the powder metal compact. Sintering temperature is usually 0.6 to 0.8 times the melting point of metal powders. For the good mixing of powders with different melting points, sintering temperature is usually above the melting point of one of minor constituents and other powders remain in a solid state. The important factors governing sintering are temperature, time and atmosphere.

Generally, scrap rates for powder metallurgy are less than 3 percent. The process has so little waste, and moreover the product is often finished when taken from the furnace. So the process is very cost effective compared to manufacturing processes that must contend with machining chips. The process is simple to complex parts that can be made to close tolerances, often eliminating machining. Production runs range in number from a few hundred to thousands of parts per hour. Conventional powder metallurgy parts are limited to parts which can be formed uniaxially.

Powder metallurgy's flexibility is becoming more pronounced with the advent of its use as a forming technique to manufacture components with very novel material combinations. The designer can adjust the chemistry of materials and other powder metallurgy characteristics, such as density, to provide a custom result, uniquely suited for the application, which is very useful to machinery manufacturing industry.

二、汉译英

金属板料成形工艺是把力施加到一块金属板上来改变它的几何形状,而不是去除任何材料的加工方法。所施加的力使金属超过其屈服强度,引起材料发生塑性变形,但不会失效。这样一来,板料可以弯曲或者拉伸成各种复杂的形状,常见的金属板材成形工艺包括弯曲成形、辊压成形、旋压成形、深冲成形、拉延成形和液压成形。

弯曲成形是在金属形成过程中,将力施加到一块金属片上,使其发生一定角度的弯曲,形成所需要的形状。在V形模具弯曲成形工艺中,凸模下行,首先接触到没有支撑的金属薄板,随着凸模继续下行,迫使材料向下,直到最终接触V形模具的底端。可以观察到,在过程开始时,板材是没有支撑的,但是当操作循环到达它的末端时,弯曲的零部件在凸模和凹模的孔隙之间得到了完全的支撑。

在对金属板料实施弯曲操作之后,材料中的残余应力会使板料发生轻微的回弹。由于存在这种弹性回复,对薄板进行精确的、一定量的过度弯曲是很有必要的,以此来获得所需的弯曲半径和弯曲角度。回弹量取决于材料本身、弯曲操作、初始弯曲角度、弯曲半径等因素。现有两种明显不同的V形弯曲方法,称作自由弯曲和校正弯曲。后者因回弹量较小,可以更好地控制角度。

辊压成形(英语为 roll forming,有时写成 rollforming),是一种将金属板料通过一系列的弯曲操作而逐步成形的金属成形工艺。这个过程是在辊压成形生产线上进行的,过程中将金属板料送入一系列辊压工位。每个工位都有轧辊,也就是辊模,位于板料两侧。其一工位的轧辊形状和尺寸可能是固定的,然而不同工位的扎辊也可能完全相同。轧辊可以在板料的上方或下方,沿两侧,或者成一个角度等。将板料强制送入每个辊压工位的轧辊,做塑性变形和弯曲处理,每个辊压工位都是得到目标零部件的完整弯曲过程的一个阶段。

辊压成形工艺可以把金属板料加工成各种横截面轮廓。开放式轮廓最为常见,封闭管状形状也可以加工。因为最后的形状是通过一系列弯曲来实现的,因此零件沿其长度方向的横截面并不要求对称。该工艺用于加工外观均匀但具有等面积复杂横截面的长零部件。

旋压有时也称旋转成形,是在金属板料一端施加力的同时,对其进行旋转形成圆柱状零部件的金属成形工艺。金属圆片高速旋转,辊轮将薄板压到一个叫作芯轴的模具上,从而形成目标零部件的形状。旋压金属零部件均呈现旋转对称、中空的形状,如圆柱、圆锥或者半球。

参 考 答 案

第一章　科技文体的特征

一、英译汉

1. 网络管理员通过直接访问和管理配置数据库文件而获得了控制能力。

2. 由于 USB 端口向外围设备提供电源，可以清除横七竖八的电源线和占空间的电源变压器。

3. 此外，DTR 集中器可以通过数据传输服务（如 ATM——异步传输模式）在局域网或广域网上实现互联。

4. 这些山丘是早期地质时代的遗留物，大部分土壤已被刮光，遭受风吹日晒。

5. NASS 起到了专用“数据移动器”的作用，在网络上传递文本、图像和视频信号等。

6. 退火处理后，样品约在 30 秒内冷却至室温。

7. 极低频电波每秒钟只能传输几个比特的信息。

8. 现有系统无法自动控制增加酸浓度。

9. 我们不应该把铸铁看作是只含有单一元素的金属，而应该注意到这种金属至少含有六种元素。

10. 它是肉眼可见的恒星之一。

二、汉译英

1. The third power of 3 is 27.

2. Traffic on each virtual circuit can be encrypted with its own distinct key.

3. The point is that each vendor has a single, much improved product set that is priced more attractively.

4. Remote access into a corporate Intranet can be achieved through a variety of methods.

5. Diamond is the crystalline form of the element carbon.

6. Almost all metals are good conductors, silver being the best of all.

7. It is a common property of all matter that it expands when heated and contracts when cooled.

8. Glass does not conduct electricity, neither does the air.

9. A whale differs from a shark in that the former is a mammal whereas the latter is a fish.

10. This current changing, the magnetic field will change as well.

第二章　科技翻译的原则与标准

一、英译汉

1. 这些分子结合在一起，在蛋白质中形成所谓的希夫碱。

2. 他推测人口在十年后可能会增加一倍。

3. 语音识别从狭窄的小市场进入广泛的大市场已有一幅清晰的指路地图。

4. 气压低，沸点就低。

5. 虽然 SVC 的单位成本高于 PVC，但如果它用得少，实际费用就低。

6. 这个加热器需要一组新的电阻丝。

7. 有限元法可以用来检测键槽和断裂面的应力分布情况。

8. 为安全起见，该元件做了绝缘处理。

9. 钠会使血压升高，因此减少盐和钠的摄入也更加有益于心血管健康。

10. 随着电力工业的发展，数字电液调节系统得到了越来越多的应用。

二、汉译英

1. Ferrous tool materials have iron as a base metal and include tool steel, alloy steel, carbon steel, and cast iron.

2. Everyone realizes to what extent the world is dependent on petroleum.

3. Even the dissolution in water cannot enable salt to change its chemical properties.

4. We now have over 2 million machine tools, and produce more than 100 million tons of oil, over 600 million tons of coal, and more than 30 million tons of steel.

5. On account of friction, the rubber sealing ring is apt to wear off.

6. The point in the Earth's crustal system where an earthquake is initiated is called the hypocenter of the earthquake.

7. Diamond is the hardest known material that can be used as a cutting tool material.

8. Virtual technology is widely utilized in various vehicle test-beds.

9. The vehicle body vibration is transferred to the road surface through the suspension and tire as well.

10. Most current technological developments in electric vehicles have centered on improving energy efficiency.

第三章　译者素养

一、英译汉

1. 不像安装在教室里由学生共享的计算机，平板电脑的特点在于人们能随身携带且能在包括家里的任何地方使用。

2. 燃料电池汽车不排放废气,产生的是脂肪酸盐和水。

3. 脱氧核糖核酸分子是由许多更小的分子组成的。

4. 美国农业部饮食模式和 DASH 饮食计划是健康的饮食模式,可提供灵活的样板,适合所有美国人维持热量限制,且能满足营养需要,降低慢性疾病风险。

5. 模拟信号由直流分量、交流分量、频率和相位四部分组成。

6. 高度完整的单晶对制造集成电路而言是绝对必要的。

7. 与传统汽车相比,混合动力汽车有减少油耗和排放的潜力。

8. 噪声指任何令人讨厌的声音。

9. 铁接触水或受潮都会生锈。

10. 炉膛内的温度并不总是处在 1000℃以上。

二、汉译英

1. Which measure of weight do pharmacists use?

2. And in this increasingly technological battlefield of rockets, anti-rocket interceptors, radars, control rooms, drones and drone hacking, it is soldiers like Idan Yahya (and whoever his counterparts on the Arab side are) who are making the most impact.

3. The average speed of all molecules remains the same, as long as the temperature is constant.

4. Not until the invention of the jet engine could airplanes travel at supersonic speeds.

5. It is the cement but not the aggregate that reacts in concrete.

6. The easiest way to achieve changes like this will be to switch to diesel.

7. Not only the transistors but also the circuit has gone wrong.

8. The electrical conductivity has great importance in selecting electrical materials.

9. The implications of the chip shortage are also being felt beyond the technology industry.

10. According to chimney effect, the higher the passage, the bigger volume of got press ventilation.

第四章　网络查询工具与方法

一、科技术语英译汉

1. 残余压应力

2. 表面等离激元

3. 法夫酵母

4. 肌肉活检

5. 链式列队

6. 包络不稳定(性)

7. 接头外径

8. 井道内电缆保护垫片

9. 光圆钢筋

10. 沉积岩

二、科技术语汉译英

1. enzyme digestion technology

2. structurally deformed coal/tectonically deformed coal

3. spatial and temporal properties/spatio-temporal properties

4. phase-selective recrystallization

5. rimmed vacuole

6. magnetic pumping

7. sectionalizing joint

8. ocean circulation

9. boring lathe

10. photoelectric functional material

第五章　增译与省译

一、英译汉

1. 本周即将开启计划的下一步,用50粒成活的种子重复做上述实验。

2. 一种元素的原子序数是其原子核中的质子数。

3. 他的证据来自对约40座火山所做的电磁调查,其中包括日本的富士山、美国的圣海伦斯火山以及玻利维亚、新西兰、菲律宾等地的其他火山。

4. 新西兰的怀特岛火山下可能蕴藏着多达140万吨的铜,而世界上最大的铜矿则蕴藏着数千万吨的铜。

5. 两种或多种物质合成一种化合物时,就产生合成反应。

6. 纺织品表面轮廓的三维成像有助于检测出传统成像技术所检测不出的瑕疵。

7. 溶剂能溶解一定量的溶质,随后就变成饱和溶液了。

8. 计算机科学的一个经典问题是确定数据应当存放在哪里,以实现最佳的读写效率。

9. 在数据输入环境中,文员会在计算机屏幕前坐上几个小时输入数据。

10. 为了减少寄生虫侵害,可以在羊身上直接喷洒杀虫剂,因此在准备过程中,羊毛加工废水中会含有这些杀虫剂残留物。

二、汉译英

1. Adequate voice fidelity requires a frequency spectrum from 300 Hz to 3300 Hz.

2. Being stable in air at ordinary temperature, mercury combines with oxygen if heated.

3. Certain bats, by receiving the echoes to the squeaks they emit, can locate and steer clear of obstacles.

4. The nutrient-rich water of Ross Sea is the most productive in the Antarctic, leading to huge plankton and krill blooms that support vast numbers of fish, seals, penguins, and whales.

5. When the explosive mixture is burned, the resulting thrust caused by the intensely hot and expanding gases can be controlled and directed to push the rocket in the required direction.

6. The lead-time problem forces facilities to be developed largely on the basis of predictions of the future, while the expense involved tends to delay them.

7. The diameter and the length of the wire are not the only factors to influence its resistance.

8. Hot-set systems produce higher strengths and age better than cold-set systems.

9. When the exploration was completed, the two astronauts on the moon would join the moonship once more.

10. Unlike satellite communications, or satcoms, standard HF transmitters and receivers can be cheap, light and compact, and require little power to operate.

第六章　拆译与合译

一、英译汉

1. 计算机是一种装置,可接收一系列含有信息的电脉冲,对这些电脉冲进行合并、整理与分析,并将它们与储存在机内的信息进行比较。

2. 越来越多易引起火灾的天气与夏季北极海冰的减少相关,这也使美国西部秋季野火日益增多。

3. 最近发现某些塑料能导电,这引起了工业界和学术界重视。因为该材料的高导电率能和昂贵金属相比拟,利用这一特性的新技术大有前途。

4. 这颗彗星由冰和尘埃组成,散发出绿色的光环,据估计直径约为 1000 米。

5. 科学家们开发出一种可诊断阿尔茨海默病的新型血液检测,无须进行昂贵的脑部影像检查或令人痛苦的腰椎穿刺。

6. 一颗新发现的彗星将在近几周内飞经地球和太阳,为 5 万年来首见,届时或可用肉眼直接观测。

7. 它们的身体是分节的,就像蚯蚓一样,有一排排的环状隔间,一遇到麻烦就能够很容易地重新长出新的头或尾巴。

8. 超级飓风把地面上的水带至大气平流层,并恶化臭氧层,从而让数以万计的生物因找不到栖息地而灭绝,直到臭氧层重新形成为止。

9. 砖电池依赖于红色颜料氧化铁,或称之为铁锈,这种颜料赋予了红砖颜色。

10. 他们把吸盘式的传感器放在鲸鱼的头上来捕捉它的脑电波活动。传感器信息显示，在有预期的噪声下，鲸鱼会降低听觉敏感度。

二、汉译英

1. It is clear that the continued development of more energy efficient technologies will be necessary.

2. Satellites give information about the surface. And scientists have launched drifting devices that measure conditions in the upper mile of water.

3. Now scientists have developed a new technique. It allows them to measure temperature changes across entire ocean basins.

4. Warming sped up the nitrogen cycle, which should have increased nitrogen's availability as plant fertilizer. But a lot of the nitrogen left the soil through run-off or uptake into the atmosphere.

5. Random numbers are hugely important for modern computing. They're used to encrypt credit card numbers and emails.

6. A lot of so-called random numbers are not truly random. They're actually what's known as "pseudo random numbers", generated by algorithms.

7. Full attention must be paid to theoretical research in the natural sciences. Negligence this respect will make it impossible to master and apply the results of advanced world science and technology and properly solve important problems in our construction.

8. The Amazon Basin has a world-famous complex ecosystem. Chinese scientists for the first time came to this region in July 2004, to make scientific probes and studies.

9. Shanghai maglev is the Shanghai maglev train line. This line, west of Shanghai Metro Line 2 of the Longyang Road Station and east of Shanghai Pudong International Airport has a total length of 29.863 km.

10. The Chinese Mars Exploration Program is the nation's first Mars exploration program. It was co-developed by the China National Space Administration and Russian Federal Space Agency with a mission of exploring Mars together.

第七章 词性转换

一、英译汉

1. 即使有防护设施，也无法保证不会发生因缺氧而致死的情况。

2. 将电路并联旨在减小电阻。

3. 黑洞就像是宇宙中巨大的吸管。

4. 所有金属都具有延展性。

5. 这些零件的比例必须准确无误。

6. 在新品种培育方面,基因突变是非常重要的。

7. 地震与地层断裂有密切的关系。

8. 每个样品均应详细标明其来源。

9. 与人类相比,计算机把检验工作做得更加细致入微,更加勤勉不倦。

10. 然而,对于这种药物应如何正确使用,人们了解得并不全面,这也说明加强对此方面的教育指导可能会取得更好的成效。

二、汉译英

1. The operation of automobiles produced overseas asks for knowledge of English.

2. We are sure that the experiment will turn out a success.

3. By mobile phone people can contact their friends far away.

4. These materials are characterized by its compactness and portability.

5. The modern weather broadcast new is highly accurate.

6. Nobody should be allowed to cause tension or armed conflicts against the interests of the people of any country.

7. The growing awareness by millions of Africans of their extremely poor and backward living conditions has prompted them to take resolute measures to create new ones.

8. The accurate regulation of the air-fuel ratio of the mixture is essential to the effective control of automotive emissions and therefore to the reduction of pollution to the atmosphere.

9. The rapid population growth worldwide, coupled with the gradual scarcity of arable land will lead agricultural products to a full bull market.

10. Rapid evaporation at the heating surface tends to make the steam wet.

第八章 反　　译

一、英译汉

1. 几乎无人会支持动物杂交的相关研究。

2. 实验数据和预期数据非常不一致。

3. 装有雷达装置的现代船只不会与其他船只相撞。

4. 在这一点上,已有人证明量子计算机能够执行一项连功能最为强大的传统超级计算机也无法完成的任务。

5. 生产抗生素灭活酶通常十分耗能。

6. 由于瓦斯爆炸实验无法控制,该研究项目被迫停止。

7. 这款辉瑞口服药能有助于防止新冠病毒感染者发展成重症,也就不用住院治疗了。

8. 这张地图显示,银河系的星系盘远非平面状,而明显是弯曲的,且各处厚度不一,距离星系中心越远处越厚。

9. 亚历山大说,除了能够更准确地了解一个地区有多少美洲狮之外,这种新的抓拍相机方法还可以用来跟踪身体一侧没有明显颜色但有其他独特特征的生物。

10. 另一方面,商业应用程序通常不是为研究而设计的,因为研究所用的数据必须可预测、收集过程透明、精细准确。有时这意味着应用程序生成的信息实际上对研究员并不那么有用。

二、汉译英

1. There is no electromagnetic wave that has a different velocity in a vacuum.

2. Electroculture is absolutely harmless to vegetables or to workers standing nearby.

3. The chemical reaction that follows is very complex, and the concrete is predictably getting harder.

4. The chamber uses an artificial light source and electric field to stimulate plant growth and prevent diseases. Operation is automatic and almost care and maintenance free.

5. Rather than sheets of ice on the surface, the water in the moon is thought to exist as water molecules bound to grains of moon dust according to scientists.

6. The DeepMind scientists developed a system that relies only on inputs from its own image sensors — and that learns autonomously and without human supervision.

7. The piezoelectric method is sensitive to pipe vibration, whereas the differential-pressure method is hardly affected by this kind of disturbance.

8. It is far from proven that freezing the nerve will result in permanent weight loss but if it does, it could have a profound effect on the lives of those who have struggled to maintain a healthy weight.

9. Two other paraplegics who received the implant are also able to move their legs, to varying degrees, and their prognosis is promising.

10. The application of liquid biopsy technology based on high-throughput sequencing has greatly increased the probability of the detection and early intervention of cancer, which is quite than what it was 10 years ago.

第九章　数字的翻译

一、英译汉

1. 白炽灯只能将自身吸收的电能的10%转化为光能。

2. 该实验室新研发的甲状腺片可使代谢率提高2倍。

3. 但是太阳黑子的数量还不到20世纪时太阳活动周期峰值的一半。

4. 这个名叫WalkSafe的程序能探测到160英尺(约48米)开外的时速为30英里(约48千米)的车辆。

5. WASP-12b的碳浓度要比通常情况高出一倍多,而且甲烷含量几乎是预期的100倍。

6. 质量大于8个太阳质量的恒星会成为中子星,大于30个太阳质量的恒星会成为黑洞。

7. 目前,从一个受试者传递给另一个受试者的指令中,有45%是正确的,比如"召集直升机"或者"前方有敌人"等。

8. 索尼公司升级后的4K数码投影机将以更快的、每秒48帧的速度来放映影片,使电影画面比目前所使用的每秒24帧的投影机放映出的更为清晰和生动。

9. 美国时间与学习中心已获得的资金超过62万美元(或:已获得资金62万多美元),通过比较核磁共振扫描来评估手镯的有效性,并制定出评价学生在课上投入程度的等级标准。

10. 该项目的第一步是研发出能够削减深空机器人探测费用的技术,将费用削减至目前高达数亿美元的航空任务成本的十分之一到百分之一。

二、汉译英

1. That is only around a fifth of the power that a pacemaker requires.

2. Bioluminescence is the largest source of light in the oceans and 90 percent of all creatures who live below about 1,500 feet are luminous.

3. The resulting films consist of a fibroin matrix filled with tiny pockets a few hundred nanometres (billionths of a metre) across.

4. An exoplanet, called WASP-12b which is about 1,200 light years from Earth and 1.4 times the mass of (or 1.4 times heavier than) Jupiter has been considered suitable for life existence.

5. The researchers used ultracold diazenylium ions (a form of nitrogen) for this test. They found that para-water reacted about 25% faster with the diazenylium than ortho-water.

6. The research group overcame the challenge by installing nearly 100 micro-cameras, each with a 14-megapixel sensor, on the outside of a small sphere about the size of a football.

7. But that is set to change with the Superbus, a 16ft-long, six-wheeled behemoth. It can carry 23 passengers and reach speeds of up 255km/h.

8. The solar car will be able to drive around 624 kilometers without stopping to recharge, and each hour in the sun will add up to nearly 9 kilometers of charge to the battery.

9. Nevertheless, with NASA estimating that there are almost one thousand asteroids over one kilometre in length and 19,500 over 100-metres in the Milky Way, scientists at the Institute of Planetary Research are trying to find a way to protect Earth.

10. The exoplanet is 22% larger and 80% more massive than Earth, making it a super-Earth. The researchers estimate that it has an average temperature of 490 degrees Fahrenheit.

第十章　主动式与被动式的翻译

一、英译汉

1. 夏季海冰大量减少时,海洋可以吸收和储存更多来自阳光的热量。

2. 3D 打印产物可用作种植产生生物能源的藻类的媒介,也可以用作研究珊瑚-藻类共生关系的工具。

3. 潮汐是由于月亮和太阳吸引海水而产生的。

4. 众所周知,空气污染颗粒物会进入人体,每年导致数百万人死亡。

5. 恐龙并非是一次性灭绝的,而是在数百万年间,由于哺乳动物之间的相互竞争而逐渐灭绝。

6. 这些耐药性基因不仅在感染了金黄色葡萄球菌的人群体内发现,也在牛和猪等家畜体内发现了。(或:不仅在感染了金黄色葡萄球菌的人群体内发现了这些耐药性基因,也在牛和猪等家畜体内发现了这些基因。)

7. 抓拍相机安装在动物经常走过的路上,比如美洲狮领土内路过的一棵树上。

8. 摄像头检测到移动物体时会触发抓拍相机,从而得到美洲狮漫步经过时的快照。

9. 这种相机甚至有红外闪光灯,无须打扰动物就可以拍摄夜间照片。

10. 预计几十年内全球气温将比工业化前升高 1.5℃。

二、汉译英

1. Computers may be classified as analog and digital.

2. The switching time of the new-type transistor is shortened three times.

3. Several elements and compounds may be extracted directly from seawater.

4. In acupuncture, needles are selected according to the individual condition and used to puncture and stimulate chosen points.

5. A fascinating portion of physics is known as modern physics, including electronics, photo-electricity, X rays, radio activity, etc.

6. Energy value of a product consists of energy values contributed by materials, labours and energies consumed during the processing, as well as the energy values provided by the transportation of energy resources and materials.

7. Over coming decades, atmospheric pollution and the greenhouse effect are expected to heat not just the air but also the surface of the oceans.

8. The face recognition technology might well have benefits, but these need to be assessed against the risks, which is why it needs to be properly and carefully regulated.

9. When regions of opposite polarity meet, prodigious amounts of pent-up magnetic energy are released as heat.

10. Methicillin resistance was thought to be tied to prescription, in part because methicillin-resistant bugs were first isolated from British hospitals just a year after the drug became available for clinical use.

第十一章　从句的翻译

一、英译汉

1. 在物理学中,加速度是指物体的速度随着时间的变化而改变的比率。

2. 人类如果非要收集碳、储存碳,就必须为此“买单”。

3. 正是这项发现让德国化学家 Johann Debereiner 怀疑是否所有的化学元素都是成族出现的。

4. 太阳给予大地热能,让植物得以生长。

5. 它们始终能探测到地表以下 2000 米以上的高导电性区域,对此最简单的解释是存在超咸的富金属卤水。

6. 对转基因食物不屑一顾的人似乎忽视了一个事实:所有粮食都属于转基因食物。野生植物与供人食用的植物之间的不同之处在于:后者经过了层层筛选和培植(也就是经过多年的突变),将优良品相进行选择并组合,使培育出的种子更大,果实更甜。

7. 这种表达并不一定精确地描述计算机内部的实际运行方式。

8. 据估计,该潜艇的核反应堆燃料能维持其全球水下航行 40 圈。

9. 现代技术给人类带来快乐的同时,也导致了人性的丧失和人类生存环境的恶化。

10. 在再热式汽轮机中,蒸汽从汽轮机的高压段回流到锅炉,再进行过热加热处理。

二、汉译英

1. In case throttling is used, the efficiency of the transmission is limited. In case adjustable pumps and motors are used, the efficiency, however, is very large.

2. This has spurred interest in new sources of the metal, most of which comes at the moment from rocks dug out of vast opencast mines that are then ground up and processed to release the copper they contain, typically about 1% of their mass.

3. Positive displacement pumps theoretically can produce the same flow at a given speed no matter what the discharge pressure.

4. They have to have a basic understanding of computers in order that people can use the advanced technology.

5. It is necessary first to convert the chemical energy into heat by combustion for the purpose that useful work from the chemical energy stored in fuels might be produced.

6. Another noticeable difference of ethanol-blended fuels under fire conditions is that when foam or water has been flowed on the burning product, the gasoline will tend to burn off first, eventually leaving the less volatile ethanol/water solution which may have no visible flame or smoke.

7. Although, in theory, SI can be used for any physical measurement, it is recognized that

some non-SI units still appear in the scientific and technical literatures.

8. Under some conditions, ethanol-blended fuels will retain certain characteristics as a gasoline-type fuel, and under others it will exhibit polar solvent-type characteristics.

9. How the system responds to such feedback can be determined through the control theory.

10. The increased fuel economy of the diesel engine over the petrol engine means that the diesel produces less carbon dioxide (CO_2) per unit distance.

第十二章　特殊结构的翻译

一、英译汉

1. 原子中心有一个原子核,环绕这个原子核旋转的是电子。

2. 不管能量的性质如何,总能量不会改变,但形式可变。

3. 石英在地球上很常见,但在火星上极为罕见,这表明形成石英的花岗岩很稀少。同时也没有证据表明,火山岩或沉积岩经受高压或高温时会产生板岩或大理石这类变质矿物。

4. 然而,眼睛的位置限制了须鲸的视野,因此它们可能没有立体视觉。

5. 吸二手烟,即吸入由燃烧的烟草从侧面飘过的烟雾或者由吸烟者呼出的烟雾,也能引发严重的健康隐患。

6. 将来某一天机器人是否能够具备和人一样好的视力还未可知。

7. 当今,主导贸易市场流通的是成品(和以前不一样,农业经济时期主要是依赖于原材料的贸易流通)。技术的进步使得产品部件的重量变轻,因此产品本身的重量也越变越轻,体积也越变越小。

8. 关注健康很重要,但也无须过分害怕疾病。

9. 结果表明,在相当长的时间内,全球平均气温都是相当稳定的,可极小的温度变化意味着环境的巨大改变。

10. 令人满意的是,生物材料的弹性模数与骨头相似。

二、汉译英

1. In the dark forest lie many lakes where a wide variety of fish are swimming.

2. Be it ever so hard, diamond can be crushed down under enormous pressure.

3. The higher an object is lifted, the greater gravitational potential energy it possesses.

4. In 1827 at Heidelberg university Wohler did discover a new element, a metal.

5. It was necessary to find a substance that could absorb neutrons so that the reaction could be stopped or kept at a level at which the release of energy did not cause the system to melt.

6. It is for these reasons that the building of such a "highway" must be done as a cooperative task force, rather than the more traditional service provider/user model that has used in the past.

7. It is generally believed that land animals evolved from sea animals.

8. It should be emphasized that wind is in a sense of a clean energy in nature.

9. It can be concluded that the Baud rate is very important to the telephone engineer, since this rate establishes the type of telecommunication channels to be used.

10. It is widely acknowledged that the automobiles and other machines have become an indispensable part of our society.

第十三章　机电工程类

一、英译汉

当今大部分发动机都配备有电子点火系统(EIS),也称晶体管点火系统(SSI)。晶体管和其他半导体设备代替了触点,起到控制初级电流通断的电子开关的作用。电子点火系统没有触点,也就无须修理、调整或更换。此外,相比有触点的点火系统,许多电子点火系统可以产生较高的次级电压。这个特点很重要,因为次级电压较高,就能使发动机在更稀薄的可燃混合气中启动。能产生较高次级电压的电子点火系统通常称为高能点火系统(HEI)。

次级电压较高,就要求火花塞线具有比触点式点火系统更高的绝缘性能,因此用于高能点火系统的火花塞线的直径大于有触点点火系统的火花塞线,解决措施是采用硅化物制线,可提供良好的绝缘特性,以承载较高的电压。

有些汽车采用电子分电器,在这种结构中,霍尔式传感器替代了原有的磁式拾波线圈。当通有电流的半导体材料薄片通过磁场,并与其垂直时,就会在导体的边缘产生电压,这个电压就称为霍尔电压。

霍尔电压与通过导体的电流和磁场强度均成正比。霍尔电压与导体切割磁场的速度大小无关。

霍尔式传感器通常贴在分电器上。在分电器内部与传感器相对的位置处装有一个小的永磁体,传感器和永磁体之间留有较小的气隙。分电器转子与分电器轴相连,转子的叶片数与发动机缸数一致。叶片之间的空间名为窗口。当分电器轴转动时,旋转叶片和窗口也会通过磁铁和传感器之间的气隙随之转动。

当气隙中没有叶片(窗口),传感器感应到永磁体产生的弱磁。只要磁场对传感器产生作用,那么传感器就会向电子控制单元发送一个低电压信号。这个电压信号使电子控制单元确保初级线圈断开,此时点火线圈初级绕组没有电流通过,然而一旦叶片转入到气隙范围,作用在传感器上的磁场效应将立即消失,霍尔电压降为零。

由于没有来自传感器的电压信号,电子控制单元关闭主回路,点火线圈初级绕组无电流流过,这将在线圈周围产生强磁场。只要叶片在永磁体和传感器之间,电流将持续通过初级绕组,叶片移出气隙的瞬间,会切断初级电路,这将导致点火线圈周围的磁场骤降,由此产生的高电压浪涌通过次级电路到达火花塞。

点火闭合时间和电流流过初级线圈时间,是由每个叶片的宽度决定的。当叶片进入气隙时,初级电流接通;而当叶片离开气隙时,初级电流切断。

霍尔式传感器摆脱了分电器对离心提前装置的依赖。电子控制单元可以使用电压信号的长度和频率,根据发动机转速的变化以实现点火提前。

二、汉译英

There has always been an incentive for transmission designers to produce a two-pedal accelerator and brake control system for a motor vehicle, as found in modern automatic transmissions. Automatic transmissions simplify car driving and for most users, offer smoother vehicle operation.

There are many different types of transmissions fitted to vehicles on our roads today, such as the manual transmission, the semi-automatic transmission and the automatic transmission (including the continuously variable transmission, CVT). Because it is difficult to achieve silent and smooth gear ratio changes with a conventional constant mesh gear train, automatic transmissions commonly adopt some sort of planetary gear arrangement, in which different gear ratios are selected by the application of multiplate clutches and band brakes which either hold or couple various members of the gear train to produce necessary speed variations. The idea of using automatic and semi-automatic gear-changing systems to lessen driver fatigue is almost as old as the car itself. Early motors were confounded by heavy and obstructive crash gearboxes, leading many car designers to investigate ways of providing a more satisfactory means of changing gear.

An automatic transmission, usually referred to as a fully automatic transmission, has therefore been defined by the Society of Automotive Engineers in America as "a transmission in which ratio changes are effected automatically without manual assist." Automatic transmissions change gear under load, i. e., the power continues to be transmitted to the driving wheels during a gear-shift operation. Unlike a semi-automatic transmission system, an automatic one completely relieves the driver of the duty of changing gear, while still allowing the driver to overdrive its normal operation if thought desirable. The power-transmission efficiency of such automatic transmissions is inherently slightly lower than that of manual and semi-automatic transmissions. However, in many applications this is compensated for by shift programs designed to keep engine operation inside the maximum economy range.

Automatic transmissions were first introduced back in the early part of last century in America and still today the automatic transmission is predominant there with very few Americans actually knowing how to drive "Stick shift". Automatic transmissions are very easy and pleasurable to drive, especially in built-up areas and towns, where the traffic is constantly stopping and starting. The only thing you have to do after engaging a gear is to press the accelerator to go and to press the brake to stop.

Load-sensitive automatic transmissions perform the drive-engagement and ratio selection (shifting) operations with no additional driver input. Every type of automatic transmissions has two sections. The front section contains the fluid coupling or torque converter and takes the place of the driver operated clutch. The rear section contains the valve body assembly and the hydraulically controlled gear units, which take the place of the manually shifted standard transmission.

The automatic transmission anticipates the engine's needs and selects gears in response to various inputs (engine vacuum, road speed, throttle position, etc.) to maintain the best application of power. The operations usually performed by the clutch and manual transmission are accomplished automatically, through the use of the fluid coupling, which allows a very slight, controlled slippage between the engine and transmission. Tiny hydraulic valves control the application of different gear ratios on demand by the driver (positions of the accelerator pedal), or in a preset response to engine conditions and road speed.

第十四章　交通运输工程类

一、英译汉

城市道路交叉口处通常不能完全满足大型车辆通行要求的标准。在数量较少、速度较低的情况下,允许它们在城市环形交叉口和信号控制交叉口占用两条或更多车道通行。交通岛、路缘线和道路附属设施必然要设置在车辆的偏移路径之外。可以通过绘制标准车辆偏移路径图检验布局,但对于更复杂的连续转向情况,推荐采用美国运输部的 TRACK 计算机程序来完成。

除了特殊装载车辆,在英国日常使用的车辆中,最长的是 18 米长的挂车组合。在空间受限的地点设计交叉口的时候,仅仅考虑其布局能满足最大车辆通行这一要求是不够的。偏移路径取决于很多因素:刚性连接还是铰接、轴距(或牵引车轴距)、前悬后悬长度、车辆宽度、挂车长度等。

在有些情况下,铰接式车辆的转弯半径以及转弯时扫过的区域要比刚性连接的长车要小。前悬后悬比较长的车辆,如 12 米长的欧洲低底板公交车,其转弯时就会产生一些特殊的问题。如果这样的车辆静止时,把转向盘打到最大转向度,当车辆启动的时候,这种车辆的后部往往是向外移动的。如果这时车辆靠近路缘,车身就容易碰撞到路边的行人。

交叉口类型的选择在有些情况下是比较简单的,而且是显而易见的。例如:

· 两个交通量比较小的居民区道路相交,采用优先通行交叉口。

· 高速公路直行段车道与其他道路相交时,采用立体交叉口。

· 交通量比较大、行人量比较大的城市十字路口,采用交通信号控制交叉口。

· 有重型货车通过的城郊双车道道路,采用常规环形交叉口。

然而,在很多情况下交叉口的选择并不是那么简单的,这时工程人员应在认真检查和分析整个实际情况后再做决策。

很多时候,人们需要升级改造现有的交叉口以使其容纳更多或不同类型的车辆,这时要考虑目前交叉口的形式和控制方法。可以参考交叉口所处的主要线路上的其他交叉口形式来确定新的或改进的交叉口形式。例如,在采用一系列连续信号控制的路口以及装有手控红绿灯的路口设计微型环岛或许是不合理的。同样,在乡村双车道道路上,本可以通过采用精心设计的优先通行交叉口或者环形交叉口来实现自由流,但如果设计成信号控制交叉口,车辆运行就存在危险。

优先控制交叉口是最简单、最常见的一种,其控制的实现要依赖于让行道路标线以及所安装的让行标志。

由于交叉路口总会聚集大量车辆,容易导致车辆冲突,因此应尽可能避免设置交叉口。在夜晚黑暗的郊区道路上,支路上的驾驶员常会误认为他们在主路上行驶。当距离十字路口较远的一侧有车辆驶来时,驾驶员的错觉会尤为明显。

二、汉译英

Sight distance is discussed in four steps: the distances required for stopping, applicable on all highways; the distances required for the passing of overtaken vehicles, applicable only on two-lane highways; the distances needed for decisions at complex locations; and the criteria for measuring these distances for use in design.

Stopping sight distance. Sight distance is the length of roadway ahead visible to the driver. The minimum sight distance available on a roadway should be sufficiently long to enable a vehicle traveling at or near the design speed to stop before reaching a stationary object in its path. Although greater length is desirable, sight distance at every point along the highway should be at least that required for a below average operator or vehicle to stop in this distance.

Stopping sight distance is the sum of two distances: the distance traversed by the vehicle from the instant the driver sights an object necessitating a stop to the instant the brakes are applied and the distance required to stop the vehicle from the brake application begins. These are referred to as brakes reaction distance and braking distance, respectively.

Design sight distance. Stopping sight distances are usually sufficient to allow reasonably competent and alert drivers to come to a hurried stop under ordinary circumstances. However, these distances are often inadequate when drivers must make complex or instantaneous decisions, when information is difficult to perceive, or when unusual maneuvers are required. It is evident that there are many locations where it would be prudent to design longer sight distances. In these circumstances, decision sight distance provides the greater length that drivers need.

Decision sight distance is the distance required for a driver to detect an unexpected or otherwise difficult-to-perceive information source or hazard in a roadway environment that may be visually cluttered, recognize the hazard or its threat potential, select an appropriate speed and path, and initiate and complete the safety maneuver safely and efficiently. Because decision sight distance gives drivers additional margin for error and affords them sufficient length to maneuver their vehicles at the same or reduced speed rather than to just stop, its values are substantially greater than stopping sight distance.

Passing sight distance for two-lane highways. Most roads and numerous streets are considered to qualify as two-lane two-way highways on which vehicles frequently overtake slower moving vehicles, the passing of which must be accomplished on lanes regularly used by opposing traffic. If passing is to be accomplished with safety, the driver should be able to see a sufficient distance ahead, clear of traffic, to complete the passing maneuver without cutting off the passed vehicle in advance of meeting an opposing vehicle appearing during the maneuver. When required, a driver can return to the right lane without passing if he sees opposing traffic is too close when the maneuver is only partially completed. Many passings are accomplished without the driver seeing a safe passing section ahead, but design based on such maneuvers does not have the desired factor of safety. Because many cautious drivers would not attempt to pass under such conditions, design on this basis would reduce the usefulness of the highway.

Sight distance for multilane highways. It is not necessary to consider passing sight distance on highways or streets that have two or more traffic lanes in each direction of travel. Passing maneuvers on multilane roadways are expected to occur within the limits of each one-way traveled way. Thus passing maneuvers that require crossing the centerline of four-lane undivided roadways or crossing the median of four-lane divided roadways are reckless and should be prohibited.

第十五章　土木工程类

一、英译汉

测量是土木工程中十分古老的领域,而且仍然是土木工程的重要组成部分。它也是一个随着数字图像和卫星定位等技术的发展而不断发生显著变化的领域。这些现代化的测量工具不仅使工程测量工作发生了根本性的变革,也对其在很多领域中的应用产生了影响。

现代测量工程包括若干个专业领域，每个领域中都需要学习大量的知识并且经过相当长时间的培训才能获得所需的专业技能。其中最主要的领域可能是平面测量，这是因为它在工程和测量实践中得到非常广泛的应用。在平面测量中，最基本的工作是距离测量、角度测量、方向测量和高程测量。人们用这些测量结果来确定位置、坡度、面积和体积等土木工程设计和施工的基本参数。平面测量是把地球表面当作一个没有曲率的平面进行测量。在一个边长为 20 千米的正方形区域中，地球曲率对位置精确度的影响可以忽略不计。然而在比较大的区域中，应该采用考虑地球曲率影响的大地测量。

大地测量学或称为高等测量学，是一个内容广泛的学科，涉及用来模拟地球尺寸和形状及重力场的数字与物理方面的知识。随着地球轨道卫星的发射，大地测量学已经成为一门真正的三维空间的学科。在大地测量中，采用了陆地和空间的大地测量技术，特别是还应用了全球定位系统（GPS）测量技术。GPS 测量技术不仅使导航技术发生了根本的变革，而且也为各类用户提供了一种高效的定位技术，其中工程界受益最多。GPS 在测量地球上地点的相对位置和绝对位置这类基本问题方面有着重要的影响，其中包括在测量速度、及时性和精确度等方面的改进和提高，可以确切地说，任何几何数据收集系统都可以在一定程度上从由 24 颗 GPS 卫星构成的卫星群中获取有用的信息。除了大地测量、平面测量、摄影测量学等方面的应用外，GPS 还应用于土木工程领域中，如交通运输（对货车和应急车辆的监控）和结构（对水坝之类结构变形的监控）等方面。

摄影测量学和遥感包含多个领域，这些领域均与从图像中获取有关物体及其周围环境的定性和定量的信息有关。这些图像可以从近距离获取，如从飞机上获取或者通过卫星获取。除了绘制大、中、小比例尺的地形图以外，许多其他方面的应用，例如资源管理、环境评价、监控等也要依靠多种类型的图像。绘制大比例尺的地形图（包括收集基础设施的数据）仍然是摄影测量学在土木工程中的主要应用方向。

二、汉译英

All structures designed to be supported by the earth, including buildings, bridges, earth fills, and earth, earth and rock, and concrete dams, consist of two parts. These are the superstructure, or upper part, and the substructure element which interfaces the superstructure and supporting ground. In the case of earth fills and dams, there is often not a clear line of demarcation between the super structure and substructure. The foundation can be defined as the substructure and that adjacent zone of soil and/or rock which will be affected by both the substructure element and its loads.

The foundation engineer is that person who by reason of experience and training can produce solutions for design problems involving this part of the engineered system. In this context, foundation engineering can be defined as the science and technique of applying the principles of soil and structural mechanics together with engineering judgment to solve the interfacing problem.

The foundation engineer is concerned directly with the structural members which affect the transfer of load from the superstructure to the soil such that the resulting soil stability and estimated deformations are tolerable. Since the design geometry and location of the substructure element often have an effect on how the soil responds, the foundation engineer must be reasonably versed in structural design.

Foundations for structures such as buildings, from the smallest residential to the tallest high-rise, and bridges are for the purpose of transmitting the superstructure loads. These loads come from column-type members with stress intensities ranging from perhaps 140 MPa for steel to 10 MPa for concrete to the supporting capacity of the soil, which is seldom over 500 kPa but more often on the order of 200 to 250 kPa.

Almost any reasonable structure can be built and safely supported if there is unlimited financing. Unfortunately, in the real situation this is seldom, if ever, the case, and the foundation engineer has the dilemma of making a decision under much less than the ideal condition. Also, even though the mistake may be buried, the results from the error are not and can show up relatively soon—and probably before any statute of limitations expires. There are reported cases where the foundation defects (such as cracked walls) have shown up years later—also cases where the defects have shown up either during construction of the superstructure or immediately thereafter.

The designer is always faced with the question of what constitutes a safe, economical design while simultaneously contending with the inevitable natural soil heterogeneity at a site. Nowadays that problem may be compounded by land scarcity requiring reclamation of areas which have been used as sanitary landfills, garbage dumps, or even hazardous waste disposal areas. Still another complicating factor is that the act of construction can alter the soil properties considerably from those used in the initial analyses/design of the foundation. These factors result in foundation design becoming difficult to quantify that two design firms might come up with completely different designs that would perform equally satisfactorily.

第十六章　材料工程类

一、英译汉

粉末冶金也称粉末成形，是一种利用金属粉末制作零部件的工艺。最初，这种工艺用来代替铸造具有较高熔点的难熔金属。随着技术的发展，低成本生产成为可能。如今，粉末冶金在金属加工领域有着很重要的地位。粉末冶金加工技术包括三步：一是混合金属或合金粉末；二是于室温条件下在模具里压实粉末；三是在可控气氛炉中烧结或加热模型，使金属微粒焊合在一起。

粉末混合

金属粉末的混合对保证产品的均匀性是很重要的。在混合之前,在混料中添加一些润滑剂,以减少磨损、减少摩擦。不同种类的粉末以正确的比例在球磨机中均匀混合。

压制

把金属粉末放在模具空腔中压制,形成贴合模具型腔轮廓的零件。用于压制零件生坯的压力范围是80~1400兆帕,该压力取决于材料以及所用粉末的特性。机械压力机常用于压制需要较低压力的物件,而液体压力机常用于压制需要较高压力的物件。

烧结

烧结是在可控环境中采用高温对生压坯加热的工艺。烧结使金属微粒更好地焊合在一起,因此提高了金属粉末压坯的强度。烧结温度通常是金属粉末熔点的0.6~0.8倍。为使不同熔点的粉末很好地混合,烧结温度常常是高于微量元素中其中一种的熔点,其余的金属粉末仍为固态。影响烧结的重要因素有温度、时间和环境。

一般来说,粉末冶金的报废率小于3%。这种工艺几乎不会产生废料,而且当产品从熔炉中送出来时产品就已经制成,因此这种工艺与其他必须产生切屑的制造工艺相比是非常经济的。利用粉末冶金工艺制造精密公差的复杂零件是很简单的,而且能够减少机械加工程序。生产运行的速度范围是每小时几百到几千个零件。传统的粉末冶金零件仅仅是单向成形的零件。

粉末冶金是制造新型组合材料零件的一种成形工艺。随着这种工艺的出现,它的灵活性显得更为突出。设计者可以调整材料的化学性质以及其他粉末冶金的特性,如密度,从而配置出一种专用的定制结果,这一点对机械制造业来说非常有用。

二、汉译英

Sheet metal forming processes are those in which force is applied to a piece of sheet metal to modify its geometry rather than remove any material. The applied force stresses the metal beyond its yield strength, causing the material to plastically deform, but not to fail. By doing so, the sheet can be bent or stretched into a variety of complex shapes. Common sheet metal forming processes include bending, roll forming, spinning, deep drawing, stretch forming and hydroforming.

Bending is a metal forming process in which a force is applied to a piece of sheet metal, causing it to bend at an angle and form a desired shape. During V-die bending, the punch slides down, coming first to a contact with unsupported sheet metal. By progressing farther down, it forces the material to follow along, until finally bottoming on the V shape of the die. As may be observed, at the beginning of this process, the sheet is unsupported, but as the operational cycle nears its end, the bent-up part becomes totally supported while retained within the space between the punch and the die.

When bending a piece of sheet metal, the residual stresses in the material will cause the sheet to springback slightly after the bending operation. Due to this elastic recovery, it is necessary to over-bend the sheet a precise amount to achieve the desired bend radius and bend angle. The amount of springback depends upon several factors, including the material, bending operation, and the initial bend angle and bend radius. There are two distinct V-bending methods, referred to as air bending and bottoming. The latter allows for more control over the angle because there is less springback.

Roll forming, sometimes spelled rollforming, is a metal forming process in which sheet metal is progressively shaped through a series of bending operations. The process is performed on a roll forming line in which the sheet metal stock is fed through a series of roll stations. Each station has a roller, referred to as a roller die, positioned on both sides of the sheet. The shape and size of the roller die may be unique to that station, or several identical roller dies may be used in different positions. The roller dies may be above or below the sheet, along the sides, at an angle, etc. As the sheet is forced through the roller dies in each roll station, it plastically deforms and bends. Each roll station performs one stage in the complete bending of the sheet to form the desired part.

The roll forming process can be used to form a sheet into a wide variety of cross-section profiles. An open profile is the most common, but a closed tube-like shape can be created as well. Because the final form is achieved through a series of bends, the part does not require a symmetric cross-section along its length. This process is for long parts with uniform but constant complex cross-sections.

Spinning, sometimes called spin forming, is a metal forming process used to form cylindrical parts by rotating a piece of sheet metal while applying forces to one side. A sheet metal disc is rotated at high speeds while rollers press the sheet against a tool, called a mandrel, to form the shape of the desired part. Spun metal parts have a rotationally symmetric, hollow shape, such as a cylinder, cone, or hemisphere.

参 考 文 献

[1] ALLEY M. The craft of scientific writing[M]. 3rd ed. New York:Springer, 2018.

[2] American Psychological Association. Publication manual of the american psychological association [M]. 7th ed. Washington D. C. :American Psychological Association Books, 2020.

[3] BARKER M, SALDANHA G. Routledge encyclopedia of translation studies[M]. New York: Routledge, 2020.

[4] BARZUN J. Simple and direct:a rhetoric for writers[M]. New York:Harper and Row Publishers, Inc. , 1975.

[5] BBC. News style guide[EB/OL]. [2023-02-05]. https://www.bbc.co.uk/newsstyleguide/all/.

[6] CATFORD J C. A linguistic theory of translation[M]. Oxford:Oxford University Press, 1965.

[7] Department of Conference Services, the United Nations. A guide to writing for the United Nations [EB/OL]. [2023-02-04]. https://digitallibrary.un.org/record/134840? ln=en.

[8] DERESPINIS F P, JENKINS H J, LAIRD A, et al. The IBM style guide: conventions for writers and editors[M]. Boston:IBM Press, 2012.

[9] Directorate General for Translation, European Commission. How to write clearly[R]. Brussels: European Commission, 2012.

[10] GARNER B. Garner's modern English usage [M]. 4th ed. Oxford: Oxford University Press, 2016.

[11] GASTEL B R. How to write and publish a scientific paper [M]. 8th ed. California: Greenwood, 2016.

[12] HUANG P P. On Chinglish in Chinese-English translation and its countermeasures: taking translation of modern Chinese prose as an example [J]. Open Access Library Journal, 2021, 8(6):1-8.

[13] JAKOBSON R. On linguistic aspects of translation[C]//VENUTI L. The translation studies reader. London/New York:Routledge, 2000.

[14] KUBLER C A. Study of Europeanized grammar in modern written Chinese[M]. Taipei: Student Book Co. , Ltd, 1985.

[15] NEWMARK P. Approaches to translation [M]. Shanghai: Shanghai Foreign Language Education Press, 2001.

[16] NIDA E A. The theory and practice of translation[M]. Leiden:E. J. Brill, 1982.

[17] NIDA E A. Toward a science of translating[M]. Leiden:E. J. Brill, 1964.

[18] NIDA E A, TABER C R. The theory and practice of translation[M]. Beijing:Foreign Language Teaching and Research Press, 2004.

[19] PELLATT V, LIU T. Thinking Chinese translation—a course in translation method:Chinese to English[M]. Oxon:Routledge, 2010.

[20] PENG J Y, XIANG X. Bidirectional remote state preparation in noisy environment assisted by weak measurement[J]. Optics Communications, 2021, 499(15):127285.

[21] PENROSE A, KATZ S. Writing in the sciences:exploring conventions of scientific discourse[M]. 3rd ed. New York:Longman, 2010.

[22] PERELMAN L C, PARADIS J, BARRETT E. The Mayfield handbook of technical & scientific writing[M]. California:Mayfield Publishing Company, 1998.

[23] REISS K. Text types, translation types and translation assessment[C]// CHESTERMAN A. Readings in translation theory. Helsinki:Oy Finn Lectura Ab, 1989.

[24] SCHULTZ D M. Eloquent science:a practical guide to becoming a better writer, speaker, and atmospheric scientist[M]. Boston:American Meteorological Society, 2009.

[25] SCHUSTER E, LEVKOWITZ H, OLIVEIRA O N. Writing scientific papers in English successfully:your complete roadmap[M]. Andover:Hyprtek, 2014.

[26] The Associated Press. The Associated Press stylebook[M/OL]. 55th ed. New York:The Associated Press, 2000[2023-02-05]. https://coppelljournalism.files.wordpress.com/2011/11/ap-stylebook.pdf.

[27] The Economist. The Economist style guide[M]. 12th ed. London:Profile Books Ltd, 2018.

[28] TYTLER A. Essay on the principles of translation[M]. Beijing:Foreign Language Teaching and Research Press, 2007.

[29] VALIELA I. Doing Science:design, analysis, and communication of scientific research[M]. Oxford:Oxford University Press, 2001.

[30] WALLWORK A. English for academic research:grammar, usage and style[M]. New York:Springer, 2015.

[31] WILLIMS J. Style:toward clarity and grace[M]. Chicago:University of Chicago Press, 1990.

[32] 陈德章. 汉英对比语言学[M]. 北京:外语教学与研究出版社,2011.

[33] 陈桂琴. 科技英语长句翻译方法例析[J]. 中国科技翻译,2005,18(3):5-7,65.

[34] 陈勇,边明远. 汽车专业英语[M]. 北京:北京理工大学出版社,2011.

[35] 程洪珍. 英汉语差异与英语长句的汉译[J]. 中国科技翻译,2003(4):21-22.

[36] 董国忠. 英语倒装句的翻译[J]. 中国科技翻译,1994(3):16-18,55.

[37] 范瑜,李国国. 科技英语文体的演变[J]. 中国翻译,2004(5):88-89.

[38] 冯庆华. 英汉翻译基础教程[M]. 北京:高等教育出版社,2008.

[39] 怒安. 傅雷谈翻译[M]. 沈阳:辽宁教育出版社,2005.
[40] 贾艳敏. 土木工程专业英语[M]. 2 版. 北京:科学出版社,2011.
[41] 李长栓,施晓菁. 理解与表达——汉英翻译案例评讲[M]. 北京:外文出版社,2012.
[42] 李长栓. 非文学翻译理论与实践[M]. 2 版. 北京:中国对外翻译出版公司,2012.
[43] 林语堂. 论翻译[C]//吴曙天. 翻译论. 上海:上海光华书局,1933.
[44] 刘宓庆. 翻译与语言哲学[J]. 外语与外语教学,1998(10):42-45.
[45] 刘宓庆. 文体与翻译[M]. 2 版. 北京:中国对外翻译出版公司,2012.
[46] 刘宓庆. 新编汉英对比与翻译[M]. 北京:中国对外翻译出版公司,2006.
[47] 刘瑛,阎昱. 材料成型及控制工程专业英语[M]. 北京:机械工业出版社,2015.
[48] 钱钟书,等. 林纾的翻译[M]. 北京:商务印书馆,1981.
[49] 秦洪武,王克非. 英汉比较与翻译[M]. 北京:外语教学与研究出版社,2010.
[50] 平卡姆. 中式英语之鉴[M]. 北京:外语教学与研究出版社,2000.
[51] 施春宏. 面向第二语言教学汉语构式研究的基本状况和研究取向[J]. 语言教学与研究,2011(6):98-108.
[52] 王力. 中国现代语法[M]. 北京:商务印书馆,1944.
[53] 王满良. 汉语无主语句的英译原则[J]. 外语教学,2000(2):66-69.
[54] 王佐良. 翻译中的文化比较[J]. 中国翻译,1984(1):2-6.
[55] 斯特伦克. 简洁的原理[M]. 余子龙,修订. 北京:人民日报出版社,2017.
[56] 邬万江,马丽丽. 交通工程专业英语[M]. 北京:机械工业出版社,2012.
[57] 谢耀基. 汉语语法欧化综述[J]. 语文研究,2001(1):17-22.
[58] 许明武,杨宏. 科技英语插入成分的翻译[J]. 华中科技大学学报(社会科学版),2005(2):120-124.
[59] 叶子南,施晓菁. 汉英翻译指要——核心概念与技巧[M]. 北京:外语教学与研究出版社,2011.
[60] 余光中. 余光中谈翻译[M]. 北京:中国对外翻译出版公司,2002.
[61] 张海涛. 英汉思维差异对翻译的影响[J]. 中国翻译,1999(1):21-23.
[62] 张美芳. 英汉翻译中的信息转换[J]. 外语教学与研究,2000(5):374-379,400.
[63] 张培基. 英汉翻译教程[M]. 修订本. 上海:上海外语教育出版社,2009.
[64] 张志公. 现代汉语(中册)[M]. 北京:人民教育出版社,1982.
[65] 郑淑明,曹慧. 卡特福德翻译转换理论在科技英语汉译中的应用[J]. 中国科技翻译,2011(4):17-20.
[66] 中国社会科学院语言研究所词典编辑室. 现代汉语词典[M]. 7 版. 北京:商务印书馆,2016.
[67] 中华人民共和国教育部语言文字信息管理司. 出版物上数字用法:GB/T 15835—2011[S]. 北京:中国标准出版社,2011.